DATE DUE

GAYLORD			PRINTED IN U.S.A.

INVISIBLE
WALLS

Further praise for *Invisible Walls*:

"The human brain today deals very imperfectly with the complex world it has inherited. The brain has built-in biases, blindnesses, deep delusions that may have made life simpler for early *Homo sapiens,* but present formidable obstacles to us, whose task is not to adapt to a deteriorating ecology, but to turn that ecology around. Peter Seidel . . . is the first to recognize this dilemma, and he makes his case with admirable verve and clarity."

—Jeremy Campbell, Washington correspondent
to the *London Evening Standard*

"This book is remarkable. . . . The author's achievement is particularly astonishing because he does not belong to those [professions who are] dealing with these problems. . . . It is amazing how, as a non-specialist, he manages to master details of social psychology or public policy that are involved in the questions he poses, and it is moving to encounter the sense of urgency and moral commitment that accompany his presentation. . . . [*Invisible Walls*] might well lead to the turnaround in public awareness and attitudes that are needed to lead mankind to its survival in the coming century."

—John H. Herz, Graduate School,
City College of New York (political science)

INVISIBLE WALLS

why we ignore the damage we inflict on the planet ...and ourselves

Peter Seidel

Foreword by Ervin Laszlo

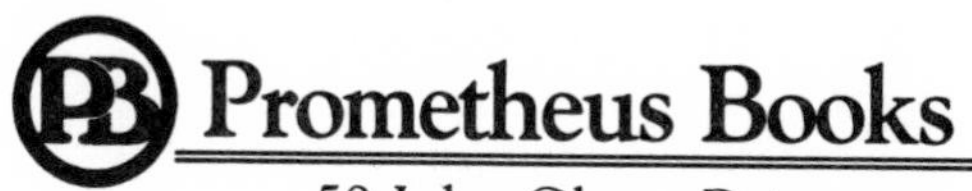

59 John Glenn Drive
Amherst, New York 14228-2197

Published 1998 by Prometheus Books

01 00 99 98 97 5 4 3 2 1

Library of Congress Cataloging-in-Publication Data

Seidel, Peter, 1926–

Invisible walls : why we ignore the damage we inflict on the planet . . . and ourselves / by Peter Seidel ; foreword by Ervin Laszlo.

p. cm.

Includes bibliographical references and index.

ISBN 1–57392–217–X (alk. paper)

1. Environmental psychology. I. Title.

BF353.S39 1998

304.2'8—dc21 98–14986

CIP

Printed in the United States of America on acid-free paper.

To

Adam

Clayton

Fang Fang

Fritz

Isabel

Jeevan

Jiegang

Joey

Molly

Sophie

Wei Wei

and Yi,

who will have to live in
the world we leave them.

How will they look back at us?

Contents

PART III: OUR ORGANIZATIONS

PART IV: SOLUTIONS

Foreword

Invisible Walls is that unique book that not only gives a meaningful overview of where we are, but also shows why we are not going where we should be. It makes practical suggestions as to how we could be moving toward where we really should be, in our joint interest.

We can no longer keep going in the direction we have been; a change has become imperative. We should heed the warning of an ancient Chinese proverb: "If we do not change the direction in which we are going, we are likely to end up precisely where we are headed." In today's world, ending up where we are headed would be disaster. There would be continued overpopulation; spreading poverty; increasing militarization; accelerating climate change; growing food and energy shortages; worsening industrial, urban, and agricultural pollution of air, water, and soil; further destruction of the ozone layer; accelerating reduction of biodiversity; continued loss of atmospheric oxygen due to deforestation as well as the poisoning of plankton in the seas; and a growing threat of large-scale catastrophes due to nuclear accident. Smaller-scale but possibly widespread disasters could also occur because of the accumulation of toxins in soil, air, and water and toxic additives in food and drink.

The direction in which we are now headed is not the only one, however. In regard to every life-threatening or depressing trend and process there are life-conserving and enhancing alternatives.

For example, we *could:*

- reduce emissions of CO_2 (carbon dioxide) and other greenhouse gases into the atmosphere;
- reforest denuded land and prevent the erosion of cultivable lands;
- reduce and clean up pollution;
- develop alternative fuels;
- do away with weapons of mass destruction and dangerous technologies;
- reduce the gap between the rich and the poor;
- inhibit conspicuous consumption;
- provide better living and working conditions for women and all the underprivileged;
- encourage a reverse flow of people from the cities to the countryside;
- retrain the un- and underemployed;
- facilitate environment-friendly business ventures;
- reallocate resources in favor of education and health care; and
- encourage smaller families

to mention the most obvious.

The constructive goals and the actions they call for are known, but unfortunately they are not acted upon. There are "invisible walls"—to use Peter Seidel's term—blocking our way. For example, in its 1995 report, the Commission on Global Governance put forward a number of achievable recommendations for enhancing international security, promoting global cooperation, managing economic interdependence, and strengthening the rule of law. The commission concluded that there is no question of *capacity* to take the required actions, only a question of the *will* to take them. In turn, the World Game Institute estimated that reallocating one-fourth of the world's military expenditures could prevent soil erosion; stop ozone depletion; stabilize population; prevent global warming and acid rain; provide clean, safe energy and water; provide shelter; eliminate illiteracy, malnutrition, and starvation; and retire the debt of the developing nations. Even small reductions in military spending could fund major health and literacy programs, create essential infrastructures, and bring marginalized people into the ambit of the modern world economy. Yet such "peace dividends," though they are much discussed, are not implemented. They do not represent true priorities on the international agenda.

Invisible walls are also blocking our way in regard to the environment. On the one hand our world is relieved of the specter of superpower confrontation while on the other it is threatened by ecological collapse, yet the world's governments spend a thousand billion dollars a year on arms and the military and—notwithstanding promises and declarations at the Rio Summit*—only a tiny fraction of that sum on maintaining a livable environment.

The right direction is by and large known; the measures it calls for are at hand. Yet these goals are not pursued; the required measures are not implemented. Why so? Seidel puts his finger on the problem. Our inherited brain structures are obsolete; our social structures are behind the times; we face an information flood producing chaos and overload; we fail to recognize the true nature of our societal and ecological systems; and our ethical and moral conceptions are narrowly focused and hence flawed. Einstein's saying comes to mind: "In this world we have changed everything but ourselves."

It is time we recognize that the changes and transformations that occur in the world cannot be mastered in our shared interest without changes in the way we perceive, think about, and "feel about" this world, including its social, political, economic, and ecological systems. We must evolve a different, more adapted, and realistic way of perceiving our states and nations, our businesses, our communities, our cultures, our life-support systems—and ourselves.

A basic change in the way we perceive ourselves and our world amounts to a change in our consciousness. All elements of consciousness enter into our view of the human and the world: perceptions and rationalizations, as well as feelings, values, and intuitions. If we are to navigate today's complex systems and bring ourselves and the next generations to a soft landing in the postmodern civilization we could

---

*Officially known as the United Nations Conference on Environmental and Development, the Rio Summit was held in June 1992 in Rio de Janiero, Brazil, to address topics devoted to developing a plan for more sustainable global development.

build in the next millennium, both the rational and the intuitive aspects of our consciousness must be evolved—the rational left hemisphere of our brain as well as the intuitive right hemisphere.

Peter Seidel's book is an enormous help in making us step back and take stock—reflecting on what is wrong and what could put it right. It is a basic tool for a deep-seated consciousness-change; for taking the first step toward responsible living and acting in our rapidly, and otherwise dangerously, changing world.

Ervin Laszlo
Fellow, The World Academy of Arts and Science;
Advisor to the Director-General of UNESCO
July 1997

Introduction

On October 7, 1988, an Eskimo hunter searching for bowhead whales off the north coast of Alaska discovered three California grey whales desperately gasping for air through a rapidly closing hole in the arctic ice. They faced certain death. This was not unusual; every year a large number of California grey whales die during their migration to Baja, California. But events took an unusual turn in 1988.

When the Eskimo returned to his base at Barrow, he reported the incident to friends in the local wildlife management office. This started a series of events which rapidly snowballed into what has been called "Operation Breakout."

Between October 8 and 28, the United States, the Soviet Union, two corporations, the environmental activist group Greenpeace, two brothers-in-law from Minnesota, and 150 journalists (including twenty-six television networks from four continents), spent $5,795,000 to free the whales and to cover the story.[1] Every day more than a billion people watched as two traditionally hostile nations joined together in an attempt to save the lives of the whales. During fifteen days of the presidential campaign that coincided with this event, the networks even cut into campaign coverage, devoting nearly 10 percent of their time to the whale story.

While this was happening, in that same three-week period, almost unnoticed, the world population increased by nearly five million; a half-million children died as a result of malnutrition; 1.5 billion tons of our planet's topsoil were washed or blown away; 2,300 square miles of tropical rain forest were destroyed; and 60 billion dollars was spent for military purposes around the world. This was not the only irony. Public interest in the welfare of the whales was intense, but short-lived. It lasted for just three weeks and involved only three whales, one of which died during the rescue. During an average three-week period in 1987, approximately two hundred whales were commercially slaughtered with little public outcry. And even in 1988, while an international moratorium was in effect, Norway, Iceland, and for "scientific purposes," Japan, continued hunting whales. After Operation Breakout was over, the USSR, one of the heroes of the event, continued on as the world's largest hunter of grey whales.

We seem to lack the ability to separate things of minor importance from the truly essential. Unlike successful plant and animal species, and our own species—until recent times—we now seem to be unable or unwilling to act in our own self-interest. This dilemma has not escaped comment.

> We know precisely and scientifically what the effects of pollution, waste of natural resources, the population explosion, the armaments race, etc., are going to be. We are told every day by countless critics citing irrefutable arguments. But neither national leaders nor society seem to be able to do anything about it.
>
> —Ludwig von Bertalanffy[2]

> We know all the things that are wrong, we know all the dangers. We have no difficulty in seeing them and describing them. We seem to have great difficulty acting upon our intuition to save ourselves, to act in a way that, in retrospect, will be seen to have been evolutionarily sound.
>
> —Jonas Salk[3]

> I think when people look back at our time, they will be amazed at one thing more than any other. It is this—that we do know more about ourselves than people did in the past, but that very little of this knowledge has been put into effect.
>
> —Doris Lessing[4]

> Our technological society displays at the same time breathtaking intelligence and abysmal lack of wisdom. That we can produce Trident submarines indicates how functionally bright we are; that we do produce them indicates how substantively stupid we are.
>
> —F. E. Trainer[5]

Our habit of ignoring the damage we inflict on our planet, and our failure to learn from it, goes back at least as far as recorded history does. Poor farming practices stripped land of its productive topsoil and seriously damaged food production in such places as ancient Greece, North Africa, China, and pre-Columbian Central America, and such practices ruined irrigation systems in Mesopotamia and the Indus Valley. Today, not having learned from these lessons, we are using farming methods that are destroying topsoil even more rapidly.

In 1896 Swedish chemist Svante Arrhenius, and in 1899 American geologist T. C. Chamberlain, unbeknownst to each other, suggested that the burning of fossil fuels might increase global temperatures by increasing the level of carbon dioxide in the atmosphere. In 1957, a study conducted by the Scripps Institute of Oceanography Research in California indicated that roughly half of the carbon dioxide released into the atmosphere stayed there. Humanity, the study noted, was "engaged in a great geophysical experiment." In 1965 a White House report to President Lyndon Johnson devoted 23 of its 291 pages to this topic.[6] It warned that by the year 2000 atmospheric carbon dioxide "may be sufficient to produce measurable and perhaps marked changes in climate, and will almost certainly cause significant changes in temperature and other properties of the stratosphere." We have not yet taken these warnings seriously.

Rachel Carson, author of *Silent Spring,* was ridiculed and threatened in 1962 when she warned that pesticides were endangering wildlife. It took many tragic experiences before actions were taken to diminish such damage. We have also responded poorly to early warnings about uncontrolled population growth, the proliferation of nuclear weapons, the dangers of nuclear power plants, rapidly declining fish catches, the deteriorating education of young people, and the destruction by air and water pollution of the historic structures in Egypt, Greece, and Europe.

This list could fill a book. Unfortunately, when problems are still ideas and theories to us—things we cannot hear, see, or feel—we tend to ignore them.

OUR DISAPPOINTING BEHAVIOR

Today's problems are so serious and so encompassing that nature can neither correct nor safely accommodate them. That leaves it up to us. But as experience demonstrates, we cannot rely on ourselves to do it either. We do not know what the world we are so rapidly changing will be like in the future. We have no plans for where we are taking it. Viewing the difficulties that we face, a rational, responsible, intelligent society would take meaningful action. We do not.

In 1957 I read a book by geochemist Harrison Brown entitled *The Challenge of Man's Future.** Brown described a number of serious threats to our civilization that were simply being ignored. They included unrestrained population growth; eventual limits to world food production; the rapid depletion of nonrenewable energy reserves and critical mineral resources; and nullifying the forces of natural biological selection. Brown concluded by describing the inevitable decline in the quality of life on a highly populated planet depleted of natural resources, and depicting the limitations this would place on individual freedom.

Many people must have read the book because its original printing in 1954 was followed by a number of reprints. The well-documented facts it presented should have aroused the public and spurred government action. This did not happen. It was as if the book had never been written.

We have all heard complaints about "those welfare recipients who spend their money as soon as they get it," leaving the future to take care of itself. This is exactly how we treat our planet and ignore our own children's future. Are we better than the welfare recipients we criticize—especially when the consequences of our own neglect are far graver? Brown's book upset me, but I, too, did nothing—at that time. It did however leave me in an awkward situation.

Several years after I had studied architecture with Mies van der Rohe, who could be considered the father of the "glass box," I went to work for a large Chicago architectural firm that was becoming famous

*(New York: Viking Press, 1954).

for designing glass-clad office buildings across the nation. There I worked on energy- and resource-wasteful buildings that were helping to bring about the very problems Brown had described. While I was studying with van der Rohe at Illinois Institute of Technology, I also studied city planning with Ludwig Hilberseimer, whom van der Rohe had brought with him from the Bauhaus in Germany. Hilberseimer's main concern was maximizing urban living conditions while consuming a minimum amount of land, resources, and energy. This impressed me.

What I had learned from Brown kept bothering me. In time, I connected the concerns he aroused in me with what I learned from Hilberseimer. There was something I could and must do.

As a member of a profession that designs the environment in which most of us live—the same environment that is largely responsible for the extravagance and waste that threatens our future—I knew that I had a responsibility. Wherever practical, I must further land and energy conservation in building and community designs. This was not possible where I was working nor would it be in other architectural offices. The most crucial task architects face is keeping busy by keeping commissions flowing into their offices. Very few of them are in a position to meaningfully influence important decisions affecting our planet.

I believed that thoughtfully conceived model communities and buildings could provide good examples for people to follow. If such models demonstrated that we could live better on less land with a lower consumption of natural resources—which would be easy to demonstrate—society might take notice and change. I was naive.

I filled the years that followed by pursuing this interest. During that time I taught; was engaged as planner for an environmentally sound, racially integrated new town; bought the land for and planned an experimental eco-community; acted as a consultant to a government agency with regard to new pedestrian-oriented communities; and constructed two small energy- and space-conserving condominiums in high-density urban areas. All of the planning projects were stopped for various reasons. The condominiums were built and featured in local newspapers and national magazines. Little interest was shown, however, in their energy- or space-conserving features.

About this time, as memories of the Arab oil boycott of the 1970s faded, public concern and government support for energy conservation evaporated. I noticed that other people and organizations with goals similar to mine were also having little effect on the public or the government—even though the popular press had been giving some coverage to environmental problems since 1969.

By now I had become aware that public insensitivity to warnings of environmental devastation did not begin with its failure to heed Harrison Brown. In 1948 ecologist Fairfield Osborn, in his popular book, *Our Plundered Planet*, warned "that if we continue to disregard nature and its principles the days of our civilization are numbered."[7] He described the problems of overcrowding, soil depletion, and forest destruction. In that same year ornithologist William Vogt's book *Road to Survival** appeared, describing similar problems. Writer and editor Clifton Fadiman noted, "*Road to Survival* should—and I think it will—arouse all Americans to a consciousness of how we are ruining the very soil beneath our feet and thereby committing suicide, not too slowly either. Let us hope it will energize a rescue squad, 140,000,000 strong."[8] Despite *Road to Surviva*l being a Book-of-the-Month-Club selection, Fadiman's hope turned out to be wishful thinking.

Presenting evidence and developing techniques is futile if they are not used. Discovering more things that threaten us, learning more about them, and devising more remedies to avoid them will do us little good if we do not put those remedies into effect. It became clear that our failure to react to planet-imperiling circumstances does not lie in our not knowing what is wrong or what to do about it, but rather in our failure to take this knowledge seriously enough to act on it. Why not, when the stakes are so high? What is it that keeps us from behaving rationally and responsibly?

WHAT WE MUST DO

If we are to move beyond our current impasse and hope to ensure a reasonable future for humanity, we must find out why we do not effectively deal with these threats. We need to gain a better understanding of this failure and then do something about it. While we need to learn more about physical problems and their solutions as well, it is essential that we recognize that the major obstacles to real progress lie deep

*(New York: William Sloane Associates, 1948).

within each one of us and in our society. In these pages, I will present a collection of thoughts leading to the idea that there are real barriers between ourselves and responsible, sensible action. As you will discover in the chapters that follow, these barriers appear in five areas: (1) In ourselves, the product of human evolutionary development; (2) in our concepts of reality and of our place in it; (3) in our beliefs; (4) in the makeup of our social structures, including governments and other organizations; and (5) in ethical systems that do not foster harmony among people or between people and nature.

ABOUT THIS BOOK

Invisible Walls is not intended to be an exhaustive study of its subject, but rather to stimulate awareness and interest. Some of the ideas here may at some point prove to be erroneous, some may even appear to contradict each other. This does not matter. A tower falls when its foundation gives way, a theory crumbles when its basic premise fails or when the links in its argument prove wrong. A beach, however, does not disappear when some sand is removed, and a collection of facts does not become entirely meaningless because some of them are incorrect. I have included many of the points that illuminate the "invisible walls" because I not only find them interesting but I believe they may make you do some thinking as well.

The amount of data on this subject is overwhelming. As with many subjects today, trying to be complete and totally current would put me into "overload" and the book would never be finished. Information on this subject is growing faster than any one person can keep up with, so I do not even pretend to try.

In these pages I offer ideas, such as those regarding human rights, that may be offensive to some people. However, if we are to deal with reality, we must face what we find.

I hope this book will stir your interest in the problem of "invisible walls," and that others, more qualified than myself, will investigate it more thoroughly.

God's Apprentice: The Source of the Problem

We live on a wonderful planet. It is bountiful, beautiful, and most important, safe—for ourselves and for the many species of plants and animals that inhabit it. It has not and will not always be that way. Once it was so hot we would have instantly burned to a cinder. Gases and toxic chemicals would have quickly asphyxiated us. We would not have had access to the chemicals necessary to sustain us. There was no intricate food chain that with the help of solar energy converted minerals into life-supporting organic compounds.

Quantities are critical here. If atmospheric oxygen were lower than 21 percent, animals could not burn enough food energy to function properly. If the oxygen content were higher, vegetation would burst into flame. Small amounts of ammonia in the atmosphere neutralize sulfuric and nitric acids there, which would kill vegetation if they fell to the earth dissolved in raindrops.

The complex balancing mechanisms we depend on evolved with life itself from the most basic building blocks of matter in hostile surroundings. Our life-sustaining environment and its support systems, such as the carbon cycle,* which circulates carbon between living

*By means of photosynthesis, plants combine carbon dioxide from the atmosphere with other chemicals to produce organic compounds. These compounds

matter and the atmosphere, are kept in balance by two forces. One is called ecology and the other is what James Lovelock calls *Gaia*. Ecology maintains population balance among the many species of plants and animals that inhabit the earth. Gaia, by interactions between life forms and the inert materials of our planet, helps maintain an environment (atmospheric composition, temperature, pressure, humidity, etc.), in which life can thrive.

Animals need free oxygen molecules in the air in order to breathe. Plants provide this oxygen by breaking down carbon dioxide and then releasing the freed oxygen back into the atmosphere. The fact that self-regulating systems and subsystems like these maintain all the conditions conducive to life on earth, that they actually developed and exist, is beyond the grasp of my mind. I can only marvel at it.

Over time, evolution produced the remarkable creatures that are us. We could adapt to changing conditions and live in a wide variety of environments—hunting, fishing, and gathering edible vegetation. Our brain evolved and expanded to accommodate the requirements of a hunting-and-gathering society. Our brain today is essentially the same as that which developed to meet those needs. It created the modern world—one far different from that with which it was well suited to deal.

Along the way, as with all creatures, our basic instincts drove us to avoid unpleasantries. Our high intelligence and flexibility gave us a considerable ability to do so. We learned to manipulate our environment in order to increase our food supply, and developed means for protecting ourselves from hostile environments, predators, bacteria, and viruses. We eliminated many natural constraints on our ecological niche, which expanded the reach of our species. At first it grew slowly, but more recently it has grown at an exponential rate, the product of our population increase multiplied by the growing demands that each of us makes on our planet.

Perhaps this course of events was unavoidable. The laws of nature are at work. Individuals work to help their species survive and enlarge its area of influence. If they are too successful at this, the balancing mechanism of the ecological system is overcome. The result is the uncontrolled multiplication of the species. Like a cancer, unless

---

become the basic building blocks of life in plants and the animals that eat them. When plants and animals die, this carbon recombines with atmospheric oxygen through decomposition and returns to the atmosphere as carbon dioxide.

stopped, it will eventually destroy or greatly damage its host, which in our case is the biosphere. Perhaps this is the inevitable end of the evolutionary process. But perhaps we, who may now have reached this critical state, can manage to practice self-restraint and in doing so continue to thrive.

When I was a child, the idea of more people and more things seemed good. People were enamored with the concepts and values expressed by General Motors' Futurama, the world of the future, exhibited at the 1939 New York World's Fair. Here everyone would find fulfillment by enjoying their personal automobile and aircraft and by driving on endless highways. One of the exhibits showed a giant machine that felled towering jungle trees with a laser beam. It was followed by a machine appearing to be five stories high and longer than three football fields which produced a perfectly graded, divided, multi-lane freeway as it moved. Growth and more of everything was good.

We seemed to have freed ourselves from the cruel whims of nature. Now it is you and I who largely determine what the earth will be like, its soil, its water, its air, and the life that lives on it. But what we are doing has not been planned. It is being done by default. It is the consequence of many little uncoordinated decisions and actions without any thought having been given to the overall outcome.

ENTER DANGER

As evolution normally works—as it did for us when our population was low—change was slow, and niches, including our own, were relatively stable. The pursuit of individual self-interest works to the benefit of each species and the system as a whole. However, when change is too great or comes too fast, danger looms.

At first this was not apparent to us. The pursuit of individual goals continued to benefit the individual. In fact most of the time it still appears to benefit the species, as side effects are either not yet noticed or ignored. By the time they are recognized, a hard-to-stop momentum toward irreversible damage may be underway.

In the United States widespread use of the automobile has led to an urban sprawl that is totally dependent on the use of motor vehicles.

We now know that the internal combustion engine is a major contributor to global warming. But because of people's love for it, the vast numbers of jobs depending on it, and the fact that our cities can no longer function without it, political leaders do not dare to suggest curtailing its use.

While evolution has given us the means and desire to change our environment, it has also left us with what were once essential, but now are often dangerous, personal and social characteristics such as competitiveness, the lust for power, self-interest (often expressed in the form of greed and selfishness), and jealousy. These characteristics exacerbate the problems caused by our influence on the earth's biosphere.

What are those problems? We are depleting our soils and poisoning the earth's water and atmosphere. We are driving many species into extinction and consequently destabilizing ecological systems. We are creating holes in the atmosphere's ozone layer, and changing the world's climate. We are exhausting precious mineral resources and leaving our descendants to deal with heretofore nonexistent radioactive substances—several of which will remain dangerous for scores of thousands of years. We threaten all life on earth with weapons we developed to make ourselves secure. In short, our technology is likely to destroy the biosphere we depend on for our very existence.

Different groups of people and individuals have their own pet worst problem, but does it matter much if one is poisoned by arsenic or cyanide? In reality, we may be done in by a combination of these dangers. For example, the food available for each of us will diminish as the world's population grows, soils deteriorate, climate change causes coastal areas to flood and rainfall to decrease in some productive agricultural areas, and ultraviolet rays harm important species in the food chain we depend on.

We are working ourselves into an ever-more-precarious position. We have developed a lifestyle and built communities that depend on a perpetually growing economy and cheap energy, water, and food supplied from distant places. The disruption of any one of these could be catastrophic. Many people choose to stake their future on luck to get us out of our present difficulties. They say conditions are not so bad and that things will take care of themselves. Do we want to depend on luck?

A WAY OUT

The human mind has gotten us into our current predicament, and it offers us the best chance for getting out of it. In spite of the complexity and magnitude of the problems we face, there are at least some things that can be done to ameliorate them. We only need to do them.

Considering the possible outcomes, the things we can do are certainly worth trying. We can work to reduce what prevents us from clearly seeing and changing our situation. And we can nurture human qualities that will guide us to act more wisely and responsibly. By so doing, we can improve our chances to survive in a satisfactory way. To do this, we must do something new for us. We must act consciously, not just as individuals, but as a species. We have been pushed into the present with its problems by our instincts and urges. We must now end this passive drifting and take control of our own destiny. To successfully do so, we need to have a better understanding of ourselves and of our society.

Part I

OUR ANCIENT BRAIN

Our minds are remarkable machines. We can recognize a face and clearly differentiate it from a billion others. What computer can accomplish that? Nevertheless, our minds are imprecise and limited. Nobel Prize-winning biochemist Albert Szent-Gyorgi has this to say: "Primarily the brain is an organ of survival. It was built by nature to search for food, shelter and the like, to gain advantage—before addressing itself to the pursuit of truth."[1] While this brain has unleashed the power of the atom, written Shakespeare's sonnets, and composed Beethoven's string quartets, it has not been as successful at creating individual happiness, managing human affairs, or responsibly controlling the technology it has developed.

We can do better, but to do so, we must better understand our brain.

1

The World as We Perceive It

Our brains have served us well until recent centuries. Now, working as they always have, they are getting us into trouble. We will start to examine why this is so by looking at the most essential part of our neural system, the part without which a brain can do nothing at all, because it would know nothing of the world around it: our system of perception.

Our perception and cognitive system includes organs such as eyes and taste buds and nerves that take messages to the brain. There are centers in the brain that interpret, filter, and combine, and memory that supplies information needed to make sense of and complete the messages delivered by the nerves. The messages that we receive would be meaningless without data stored in memory that tells us that the lines, shapes, and textures we see are a house, and then goes on to say a lot more: It is our house and it is made of wood.

FILTERING

A vast amount of information describing our environment surrounds us. The information available is far more than our brains can handle.

If we were able to hear all of the sounds occurring around us, from low pitches to ultra-high ones, for example, we would not be able to understand what someone was telling us. A cacophony of sound would surround us.

Our environment has no color, sound, smell, or temperature as we know it. Instead our surroundings bombard us with electromagnetic waves (light) and other radiation; air vibrations (sound); aerosols and gases (odors), and objects that can be felt and moved to reveal information about their size, weight, texture, hardness, and temperature. Our perceptual system, consisting of sense organs and neural processing centers, selects a small portion of this, and in our mind, transposes it into images of colors, sounds, and objects, creating a picture of reality that we can understand. It is an abstraction of a very small part of the reality around us.

To meet their needs, other species observe the world differently than we do. They are sensitive to somewhat different wave bands of light and sound. Evolution gives creatures the means to perceive what is important for their survival. Assimilating more data than is needed would cause neurological overload, slow thinking, and confusion, and could therefore be very dangerous.

"People experience only about one trillionth of outside events."[1] When you look at your finger, your eyes are incapable of observing individual cells, the molecules of which they are made, or the voids that make up almost all of the volume of the atoms that compose them. You can perceive only a small band of the radiation reflected by the surface of your skin, and nothing behind it.

We have inherited cognitive mechanisms with data filters that were suited for the simple lifestyle and physical environment of our predecessors. While it told them all they needed to know, our situation is different. We have produced materials we cannot see, such as chemical toxins and radioactive materials, that are dangerous to ourselves and our environment. We do things we do not observe, such as increasing atmospheric levels of sulfur dioxide, which ultimately causes lakes and forests to die.

UNNOTICED VOIDS

When we look at something, we do not receive the complete detailed picture of it that our mind tells us we are getting. Only a very small area at the center of our visual field reveals details. At one hundred feet, roughly an area of one inch is sharp, the rest is vague. Swift eye movements, memory, and tricks of the mind give us the misleading impression that we are getting a photographic picture all at once.

We all have a blind spot (about six degrees wide) near the center of the visual field of each eye where the optic nerve attaches to the retina. We do not notice it, as our mind fills in what we do not see. To demonstrate, hold this page eight to ten inches from one eye, and keep the other closed. While keeping your open eye fixed on the **X**, move the page back and forth until one **O** disappears. If you repeat this process with your other eye, the other **O** will disappear.

O **X** **O**

We also observe poorly for reasons that are not physical. We miss parts of a lecture or a movie because our mind wanders or rearranges ideas. Some critical parts of an argument can easily be missed without our noticing it. Most of us are aware of this difficulty, but sometimes we refuse to recognize our failure. We are unaware of gaps and insist that our perceptions are correct and complete. The incompleteness and inaccuracy of our observations is well illustrated by the varying testimony of witnesses at a trial. Each may, in all sincerity, swear to different versions of the same event.

In 1921 Nicola Sacco and Bartolomeo Vanzetti were sentenced and put to death for murdering a factory cashier and his bodyguard in the course of a robbery in Massachusetts. Immediately after the crime, witnesses gave conflicting descriptions of the perpetrators. One even said that one of the assailants had very light hair, like a Swede, rather than dark like both Sacco and Vanzetti had. Others admitted that they had not seen clearly enough to identify anyone. Over a year later, during the trial, these same witnesses identified both Sacco and Vanzetti as being the persons who had committed the crime.

The trial drew national attention and protests. Psychologists became interested in the inconsistencies and changes in the accounts of the witnesses. Some conducted studies of this phenomenon. Years later psychologist Elizabeth F. Loftus reviewed these studies, conducted experiments of her own, and published her fascinating book, *Eyewitness Testimony*, documenting the frequency of discrepancies like this.[2] Observations of the accuracy of my own perceptions and memory sadly coincide with its disturbing revelations.

We often fail to notice what does not interest us, what we do not see as important, or what we are not prepared to see. Ulrich Neisser, formerly professor of psychology at Emory University, cites an experiment where subjects watched a basketball game on a videotape. They were told to watch the ball carefully. During the game a a woman carrying an umbrella walked across the court. When asked if anything unusual had happened, only 21 percent of the subjects reported having noticed her. In the United States, many bicycle riders are hit by motorists because the motorists, however alert they may be, are looking for cars, not bicycles.

My freehand drawing teacher at the University of Colorado would occasionally ask his students to look at something for a few minutes, and then look away and draw it. This was a humiliating experience. Try it. It taught us that we see much less and more poorly than we think we do. For instance, if the object was a shoe, I was quite sure that what I had in my mind was a fairly complete and accurate representation of it. Not so. When I started to draw, I soon got stuck, for I did not know how the leather was joined at the back, how many holes there were for the laces, and so on. I soon learned there were extensive gaps in my observations, and what I *was* sure of was often incorrect. I think society would benefit if we were all required to repeat this exercise a number of times during our education. It would make us more humble about our power of observation.

Many of the problems of our time develop because we perceive incomplete, overly simplified pictures of reality, and then, believing them to be accurate, interact based on these misconceptions.

WE ONLY NOTICE CHANGE

The only thing we perceive is change. Our perceptual system only responds to differences, such as a change in color or intensity in our visual field; a change in volume or pitch in sound; or a difference in temperature. If there is no change, we notice nothing; if change is too gradual or too small, it will pass unnoticed.

If a frog sits in a pot of water that is heated slowly, it will stay there until it boils to death.[3] No sudden change occurred to alarm it. Encounters with pots of slowly heated water are infrequent occurrences in the lives of frogs. They do not need to perceive gradual changes in water temperature.

The part of the human brain known as the hypothalamus helps keep anxiety levels low so that we are able to think clearly. Changes arouse it so that indications of danger or opportunity to meet a need can be passed on to higher levels of the brain to be acted upon. However, change that is slow or that did not affect our survival in the past does not arouse the hypothalamus. This is dangerous to us now.

Seemingly Innocent Gradual Change

Today many seemingly gradual changes pose serious threats to our future but do not arouse our concern. In fact, we have grown so used to gradual change that should it stop, we would take notice and worry. We become alarmed when the Gross Domestic Product or a company in which we own stock ceases to grow. Businesspeople are disturbed when a community's population does not increase or when new home construction on the city's periphery stops. People who foster such change usually see themselves as being conservative and those who want to block it as "troublemakers." We do not notice that, day by day, the gradual changes we accept as normal and necessary to maintain our lifestyle are bringing about radical changes that are damaging the biosphere and undermining our civilization. To us, each day seems the same, but it is not. Each day there are more people on the earth to feed and less clean water and cropland to provide for them.

Exponential Change

The most dangerous form of change is that which increases exponentially—i.e., the rate of change stays the same, but the amount of change grows larger as change proceeds over a broader area. Not only do we fail to notice this in things like population growth, but our minds have difficulty understanding the consequences of it. Simple logic would tell us the terrible consequences, but we do not go through a logical process.

In 1968 Stanford University biologist Paul Ehrlich described a hypothetical situation that would exist in 900 years if the world's population continued to double every thirty-five years.[4] There would then be 60,000,000,000,000,000 (60×10^{15}) people on the earth. That is one hundred people for every square yard of land and sea, including the North and South Poles. High school mathematics can reveal this, but we don't bother to investigate the consequences of unceasing growth.

Change Slips Up on Us

Because many changes are slow, we fail to notice that many things were once more pleasant. People expect to stay in a high-rise hotel when they vacation at the crowded beach. They do not think of how wonderful it must have been to stay in a quiet seaside cottage.

We do not develop the concept of object permanence—the ability to remember the existence or location of objects that have been removed from sight—until we are about eight months old.[5] As adults, with some effort we can remember things as they used to be. In our daily routines, however, we forget that once people did not have to endure many of the negative aspects of progress we now accept as normal. As our environment slowly deteriorates, it does not occur to us that once people would have abhorred what we put up with today, such as the long commutes on congested, ugly highways, or being unable to eat the contaminated fish we have caught nearby.

It has taken decades of the unceasing efforts of whistle-blowers for sizable numbers of people to become concerned about environmental degradation. The public was not aroused by noticing the changes, but rather by being told about them. When we do become aware of a slow change like that occurring in nature, it is only noted on an intellectual level. This is very weak when compared to the direct response evoked by rapid change. For example, if your hand inadvertently brushes against a hot burner, instinct causes you to jerk back, away from danger. If we only noticed this change on an intellectual level, then a serious burn could be incurred before any action was taken. Concerned individuals encounter opposition when trying to generate public interest in the gradual destruction of large areas of a nation's land by strip-mining, clear cutting (removing all the trees from a sizeable section of forest), harmful farming methods, or overgrazing. In contrast, a military attack followed by the occupation of a few hundred square feet of a nation would draw a quick and unified response from citizens.

There is an odd twist in how we relate to slow change. We are like the frog in the pot of heating water, but with an added feature. Often, because it makes us feel good, it is we who turn up the heat. We like what we do despite the damage which results, so we keep doing it. We want more automobiles, fast food, and throwaways. We like things as they are, with, of course, the gradual introduction of still more "improvements"—which means still more consumption and waste.

LOST IN A SEA OF NUMBERS

In prehistoric times we needed only to be aware and have a limited understanding of what was around us. It was not necessary to comprehend very large or extremely small numbers, distances, objects, or periods of time. People did not need to know the distance to the moon or the size of bacteria.

Today, very large and small numbers are part of our lives. A friend of mine who is studying the origins of galaxies works with them mathematically, but admits that he cannot comprehend the dimensions and times involved. Although I know approximately how many people live in Chicago, or how far it is from there to San Diego, these are merely numbers to me. While we now are making great changes to the earth's biosphere, our senses do not comprehend what we are really doing. We must rely on instruments and statistics help us to do that. But with few exceptions, this knowledge does not have the impact of personal experience.

DISTORTIONS OF OUR PERCEPTUAL SYSTEM

In our complex society, the intensity of our perception of something is not in direct proportion to the importance of the thing itself. Consequently, we may be fearful of being mugged on a city street, but oblivious to the dangers of uncontrolled population growth.

Our perceptions of light and sound are more logarithmic than arithmetic in magnitude. When a light or sound intensity increases by a factor of ten, it seems to us that it is closer to doubling, and when it increases by a factor of one hundred, it appears to have tripled. While this enables us to perceive signals over a wide range of intensities and to recognize changes in intensity, it confuses us about actual magnitudes. Time also distorts our perception. Because we become acclimated to change, over time a loud sound does not seem as loud and a bright light ceases to seem as bright. A similar desensitizing can take place over a longer period of time when no obvious consequences, such as instant climate change, follow repeated activities, such as the emission of carbon dioxide.

Often we can function well in spite of our limited comprehension of reality. However, when accurate assessments are needed, we can make serious mistakes. When our feelings do not reinforce what we observe, it is hard for us to comprehend the seriousness of environmental dangers, a war, or a deteriorating public school system.

Perspective makes a far-away object appear smaller than it would appear if it were close. I like to climb mountains, for example, and often have the following experience: After much effort I reach a point that from the bottom appeared to be halfway to the top. On reaching it, however, I find that most of the mountain is still above me. The same holds true on the way down. If I pick out a point that looks like it is halfway to the bottom, on reaching it, I find out that most of the mountain is still below me. What appeared to be the midpoint was not that at all because what lay beyond it was far greater than the part that was close to me—even when I tried to compensate for the perspective effect.

Often we compensate for this misleading appearance intellectually, but sometimes we are mistaken. An object or the source of a noise loses its importance to us as it recedes into the distance. If it is very far away we may not see or hear it at all.

We are very aware of the present and recent past. The more time elapses after an event, however, and the weaker its impact on us, the less real it seems and the more easily it is forgotten. Even though we know it is not true, people and things that are far from us in time, distance, or relationship seem less real and less important than those that are close. An automobile bearing down on me is very real and needs my attention whereas one that is two miles away does not.

Reasons for and Dangers of Distortion

Once these subjective distortions served a useful purpose, and often they still do. However, they can lead to serious problems. In earlier times, we could only influence and be influenced by that which was close to us in scale, distance, and time. Our perceptual system did not feed us data that would distract us from focusing on survival. That system leaves us with an incomplete picture of the greatly expanded reality of which we are now a functioning part. As a consequence, our reactions to it are often inappropriate.

We now live under conditions where our feelings about an event can be totally unrelated to its importance or to our actual involvement with it. Most of us are unmoved or even pleased when the U.S. government cuts foreign aid that has been partly used to feed malnourished children. On the other hand, some of these same people attending a movie may be greatly moved by injury to a fictitious dog. The state of the economy, which affects our pocketbook, appears more important to us than issues involving global survival, which we cannot feel. Today, we must often make decisions about such situations—without having the mental equipment to evaluate their true importance to us. (How to improve our decision making is described in chapters 13 and 15.)

OUR SENSE OF TIME

Humans have a poor comprehension of time: We only think of it in terms of our own lifespan and tasks that we do. Our grasp of it is best when measured in minutes or hours; we have increasing difficulty relating to and comprehending periods that are much shorter or longer.

What may seem like very slow change for me, in my short life, can be incredibly fast for evolution. We are now changing the environment so rapidly that many species cannot adapt fast enough to be safe. Harvard entomologist Edward O. Wilson estimates that the "number of species on Earth is being reduced by a rate from 100 to 1,000 times higher than in prehuman times."[6] When change happens too fast, neither evolution nor feedback systems* are able to restore stability rapidly enough to prevent a dangerous state of accelerating imbalance.

Evolution created us to further the long-term well-being of our species indirectly by reacting to short-term urges and needs. There is no foresight in evolution; all species evolve by interacting with existing conditions. Squirrels that store food for winter do not do so out of foresight, but rather because those that did so survived and produced others with this trait. Because ecological niches changed slowly, foresight was not normally needed. Abrupt dramatic changes could bring about the extinction of a species, but then other creatures would evolve to fill the niche.

Our inborn emphasis on what is close by leaves us with little interest in the future. Today we must not just concern ourselves with, but also plan for the future, though our intrinsic motivation to do so is weak. This varies from individual to individual, with some, such as alcoholics and criminally delinquent children, not seeming to look forward at all.

---

*Feedback occurs when a portion of the output of a process or system is returned as input. By so doing, a control is brought into the process or system. Negative feedback occurs when this process works to maintain a stable condition. For example, when a furnace heats the air in a home, some of this heated air affects the thermostat which turns off the furnace. When the temperature drops, the thermostat turns the furnace back on. This process maintains the temperature at a relatively stable level. Positive feedback, on the other hand, destabilizes: The furnace would go on when the temperature rises and off when it drops, increasing the already present disparity.

We are usually immersed in events that demand immediate attention. Many people know that it is wise to save for their children's education, for a house, for illness, but we are unlikely to prepare for things that have not happened before or which are remote in space or time. Despite warnings, people will build houses in floodplains if they have not experienced floods before. We are not ready to accept the idea that we can overpopulate our planet, change its climate, or destroy most life on it with a nuclear catastrophe. We are simply not in the habit of locking the barn door until after the horse has been stolen.

WHAT WE DON'T SEE

Before, and during the period that life has existed, chemical and geological processes formed an environment that was relatively free of radiation and other things that are harmful to living cells. Toxins such as heavy metals were for the most part left in deposits or in stable compounds safely away from our food and water. Harmful ultraviolet light and cosmic rays were filtered by the atmosphere, but radiation necessary for photosynthesis reached the earth's surface. The earth's atmosphere and life forms influenced each other to produce an environment that was safe and nourishing for life as we now know it.

We have altered this environment so that it is no longer free of the toxins and radiation that can harm life. It is rapidly losing the topsoil and other natural resources (located out of view of most people) that are needed to sustain the rapidly growing human population, which demands more material goods and produces increasing amounts of pollution.

As we drive our cars, we release oxides and other chemicals that are harmful to vegetation and animal life into the air. We know that piping automobile exhaust containing carbon monoxide into a car window is a means for committing suicide, yet, it does not bother us that thousands of cars spew this gas into city air every day. We rarely see or smell these pollutants, but when we do, we are so used to them that we pay no attention. Intellectually, scientists note this damage, but decision makers and the public, unable to feel it, are slow to take corrective measures, which, when enacted, often fall short of what is needed.

In the introduction, I described how warnings about global warming had been made a number of times since 1896. Although this information should have aroused people's concern, it was ignored and largely forgotten until recently when scientists reassessed the subject. At the 1992 United Nations Earth Summit in Rio de Janeiro, then U.S. President George Bush, representing a public apparently unable to weigh correctly the seriousness and urgency of the situation against short-term interests, greatly weakened the international agreement to reduce climate change by insisting that targets and timetables for cutting greenhouse gases be removed from the treaty.

Life has always depended on things that are distant, such as oxygen-producing plants in oceans and forests. It has always relied on unseen feedback systems that maintain ecological, chemical, and climatic balances. While people have always influenced things they did not see, in the past, this really did not matter; humans were few and the earth's self-regulating systems maintained stability. Today, the exploding human population, with billions of people impacting those unseen systems, has disrupted many feedback systems and thrown our biosphere into a rapidly increasing state of instability.

We are now living in an environment that is rapidly going out of balance, but we have no natural mechanism that adequately alarms us about the dangers this entails. Our mental images of reality omit much of the data we need in order to understand this situation and deal with it safely.

OUR INNER MODELS OF THE WORLD

Being aware of food caught in its tentacles may be all the information a simple creature such as a sea anemone needs. The signal the food sends to the anemone directly causes the anemone to guide the meal into its mouth.

Higher forms of animal life need much more detailed information about the world around them. A fox, for example, needs to know the territory where it lives, including what is good to eat and where to find it and what its own wants, capabilities, and limitations are. It needs an inner picture that provides all this. This picture tells the fox that over the next hill there is a pond where ducks tend to rest at night and that these ducks are tasty and easy to catch.

As I explained earlier, the world we (and foxes) perceive is not one of colors, sounds, and odors, but of physical phenomena such as electromagnetic waves (light) and molecular vibrations (heat). These mean nothing to us, or foxes, so we translate these perceptions into more comprehensible images. We combine new perceptions with data already stored in our memory to make mental models of the world around us. These models are also stored in memory and are continually modified as more information comes in. They are what we envision ourselves to be interacting with. They are what our brain uses to help us "see" and think about the world around us. Something in us tells us that these very limited models of the world are reality itself.

Faulty Models Mislead

Our models are far from perfect, however. They are incomplete, often inaccurate or dead wrong, and are sometimes inappropriately used. This, of course, affects how we and society think and act.

Sometimes the sum of all the models each of us carries around inside our head is called a "world model," but I prefer to think of many models or images because, as I will explain in the next chapter, our minds appear to consist of many loosely linked or unrelated parts that are also often poorly connected. Although all things in the world are in some way related, we often have difficulty establishing links between our models or even between the parts within them—if we link them at all.

The potential buyers of condominiums I have built illustrate this weakness well. Most people were very concerned about the parts—the bathroom, the range, the sliding glass door, the latest accessory, or what have you. Few noticed relationships such as the placement of the sliding door and its proportions, good cross ventilation, the thoughtfully conceived floor plan, and harmonious materials. The parts meant something as symbols, but few prospective buyers were able to evaluate the units as wholes. Perhaps that explains why many builders' houses are poorly related collections of trendy symbols. But there are graver aspects of this problem.

We fail to relate air pollution to urban sprawl, which requires people to do an enormous amount of driving to meet their everyday needs. To alleviate the pollution, we enact laws that mandate automobile exhaust control devices that actually increase fuel consumption, and we promote mass transit—which does not work in low-density suburban areas. Meanwhile, we allow the sprawl to proliferate, forcing people to drive farther and farther, creating still more pollution.

Individual economists may understand that the carbon dioxide buildup in the atmosphere is dangerously affecting the earth's climate, yet they totally ignore this fact when they propose solutions for economic problems.* When experts do not connect primary causes with problems, the rest of us assume that it is all right to ignore such connections. More likely, we don't even think about it.

We are interested in how much a person knows, but show little concern for her ability to make any but the most obvious connections. People who know a lot and can present themselves well often gain responsible positions in society—even though they may be unaware of important relationships between things. This is serious when these people make far-reaching decisions that affect us all.

FACTORS THAT DISTORT OUR MODELS

I can only observe the world from my own specific period of time, geographical location, social position, belief system, and related perspectives. Each of us is a citizen of a specific nation and belongs to religious, occupational, economic, and special interest groups. These viewpoints affect our images of the world. A contemporary Wall Street investment banker has very different models of "reality" than I do or a nineteenth-century plantation slave would. Some people see a redwood tree as majesty, others as part of an ecological system, and others as potential dollars.

Various factors determine what we select or reject for incorporation into our models. Our primitive instincts, what we already know, our interests, concerns, beliefs, misconceptions, prejudices, intelligence, and peer pressure affect our models. They subconsciously

---

*Perhaps this can be explained by the fact that governments and businesses demand short-term projections and show little interest in the more distant future.

influence our filtering mechanism and distort data. This not only leaves us with an incomplete picture of reality, but skews and warps that picture, introducing false information.

Sometimes we are mistaken in what we observe and on occasion we are inadvertently or deliberately fed lies—sometimes by ourselves. Our models are based on what information gets through our filters, the form it takes, and how our minds work with the data and the value placed on it. For some people baseball cards or Elvis Presley memorabilia have value, coastal wetlands do not. Such models then become our concept of reality.

We each know little of the world. We see it subjectively from our limited perspective, but in our mind we have the impression that what we know is correct, and important. We therefore run into conflicts with each other, each of us having models that we "know" are right.

MODELS DISCONNECTED FROM REALITY

Until very recently, even city dwellers felt a relationship to nature. The land around their cities provided most of their food, which varied by the season. These environs were also where many of their families were rooted and where they often went for recreation. When they traveled elsewhere, they passed through the nearby countryside and breathed its essence and heard its sounds. Most modern city dwellers, however, have little relationship to nature or their surroundings. Many grew up a thousand miles away. Their food comes packaged, from unknown distant locations, having nothing to do with the season. They get their milk in cartons from supermarkets rather than from cows. They vacation and do business far away—in places they have come to know better than the land around their city or their own backyards. If they do pass through their local countryside, they do so rapidly, in sealed, air-conditioned compartments distracted from the reality around them by audio entertainment.

Urbanization and prosperity have dulled our awareness of and feeling for the soil. We are as dependent on it as we ever were, and although we know this to be true, we forget. We replace our innate awareness of being dependent on nature with the belief that our well-being relies most heavily on monetary, political, and military systems and technology. This significantly colors our model of the world we live in.

The most important features on most maps and globes are human-made boundaries and cities—even the edges of oceans are considered political boundaries. Mountains, deserts, drainage basins, eco-communities, vegetation, and soil types are relegated to special maps that are of little interest to most people. Maps show what is important to us—and what is not. They show that we see the world primarily in political and economic terms. The world is divided into what were practical, defensible administrative areas before the age of airplanes, intercontinental ballistic missiles, and electronic communications. Astronauts and all creatures other than humans see a very different world. They see an undivided world where national boundaries don't mean a thing, but where natural features mean everything. Our view of the world is myopic, arrogant, unreal, and dangerous.

RELATIONSHIPS TO NATURE

Indigenous people had a much more realistic relationship with nature than we have. We have become estranged from it. Our artificial lifestyle and environments separate us from food production and waste disposal—and from natural sights, sounds, and smells. Professional mind manipulators have replaced the messages we formerly received from nature with trivial information about deodorants and beers. Our interests, values, and goals have shifted from the natural world to an artificial one, from reality to what is often an illusion.

While many Native Americans saw themselves as part of nature, most modern Europeans and North Americans see nature as a resource to be exploited. We seem oblivious to what nature does for us: It maintains the very narrow range of temperature, pressure, and humidity that is necessary for our existence. It provides us with the chemicals and radiation we need and shields us from those that can destroy us. It nourishes us by means of a food chain originating with photosynthesis and the simplest forms of life. The conditions we need to live are very limited, rare, improbable, and delicate, a fact we do not understand. We are now merrily destroying our own life support system without seeming to comprehend or be concerned about it.

OUTDATED MODELS

We base the way we interact and communicate with other people and nature on our models, which can vary greatly from person to person. Many of our models are hopelessly outdated. In spite of our continuing efforts to update them, society's models for the most part lag far behind those of its more forward-looking members. Anthropologists call this "culture lag." Although many antiquated mental models have gradually been discarded in recent years, others are still used for making decisions and planning our future. Many people still envision the earth to be a never-ending source of natural resources and a bottomless pit for trash. Others see the nationalism that developed at the end of the eighteenth century as the natural and ultimate way for the world to be organized—in spite of recent developments in communications, transportation, and weaponry. Our models are influenced by our beliefs, as we shall see in chapter 7.

Differences can lead to misunderstandings, conflicts, and mistakes. When these models are skewed or wrong, the consequences can be anything from comical to tragic—including the destruction of most life on earth.

2

The Limitations of Our Brain

People who use trucks are usually very aware of just what they can and cannot do. A truck driver knows how to make his vehicle perform efficiently without overtaxing it. But humans are painfully ignorant of a much more important piece of equipment—our brain. We rarely use our brain's abilities effectively, and we rely on it to do things it only does poorly. A better understanding of the potential and limitations of our mental equipment would greatly benefit ourselves and society as well.

What a truck can do and how much it can carry are determined by its design, construction, and size. It is the same with our brain. Let's begin by looking at its size.

As human evolution and a comparison to other primates reveal, a large brain has a greater thinking capacity than a small one. Although we are smarter than chimpanzees, we are still limited in our thinking ability by the size of our brain. Theoretically, if it were larger, we would be able to remember more things and remember them in greater detail. Thinking might be more thorough, but it would also be more complicated. Because our brain is prone to error, this greater complexity would increase the likelihood of error. Longer neurons (impulse-conducting cells that comprise the brain, spinal column, and nerves) with more connections would also slow down the thinking process. A heavier head would be a greater burden for a pregnant mother to carry and she would also need a larger birth canal. So the size of our brain is a compromise between capability and practicality. As psychologist Robert Ornstein has said, "with a complete mental system it might take us fifty years to decide if something was an emergency, and we might have a head so large we couldn't run away, anyway."[1]

The human brain's basic structure comes from early periods in evolution. Without wiping out its ties to a primitive past, the brain was added to and reformed to meet the needs of ever higher forms of life, and eventually those of a human hunting-and-gathering society. The result is a trade-off between constraints imposed by the past and the needs of society. To achieve this, the brain utilizes a variety of remarkable techniques which, in turn, impose limitations on how we think.

HOW OUR MINDS WORK

Computers can handle large quantities of data rapidly and precisely. They are capable of logical processes, precise data storage and recall, sorting, producing graphic images, and other operations. In order to accomplish these things, they must be presented with material that is unambiguous and complete, and the problems they are expected to solve must have straightforward solutions.

The human brain developed in an environment that was not so orderly. The information it had to handle was often incomplete, contradictory, and ambiguous. Some words had multiple meanings. People were sometimes purposely presented with data that was false. Tasks too were often ambiguous, open-ended, and messy. Even now, humans can consciously only think in one dimension, i.e., sequentially, one thing at a time. With this limitation they have to deal with a four-dimensional world (time is included here as a dimension).

Solutions to problems are required even though their existence is not certain, and the time to produce them might be inadequate. Results are more important than accuracy or truth. Speed at the expense of accuracy (accepting occasional mistakes), relying primarily on data association instead of logic, and other techniques help us cope with our physical and structural limitations. Our brains have developed ways to deal with the uncertainties around us and because they can do so, we as a species survived, flourished, and developed the complex world we have today. However, these techniques, no matter how ingenious, are not enough to deal safely with the problems we now face.

INFORMATION HANDLING

As I explained in chapter 1, our sense organs perceive only a very small part of the information available about the world around us, and what we do perceive is still too much for our brain to store or manipulate. We cannot, and fortunately need not, process and be able to recall all of the information that enters our sense organs.

When someone talks to me, I do not need to be aware of all of the sound waves she produces, although I will hear and notice enough for voice recognition and will pick up on emphasis and mood. What I do need to take note of are words and their meanings, but unless I have an unusual memory, I will not remember most of these either. What I hopefully will remember are the ideas that were meant to be conveyed. However, over time, I will not be able to recall all of them either, unless they have special significance for me. This is a brief example of the process of filtering, which does have an advantage: If our memory stored much of the irrelevant information our senses perceive, recall and thinking would be more difficult and slower than they now often are.[2] If we were constantly aware of everything that needed to be dealt with, we would be so overloaded with information and stimuli that we simply could not function. Even with the filtering of input, we are still burdened with huge amounts of data, much of it superfluous, and much of it hard to correlate and utilize.

So we select, eliminate, simplify, and arrange data and its relationships in a manner that enables us to handle it in our imprecise way. We discard what does not interest us, give weight to what does, and place values on what we see as meaningful. Sometimes we confuse things that are of little importance for us with that which is critical to our survival.

Jeremy Campbell, author of *The Improbable Machine*, writes: "the mind tends to reshape the information it is given into a form that is restricted in certain ways, less logical, more prone to error, wiping out fine distinctions. By doing this the mind can avoid the disaster of combinatorial explosion [a head-on crash between the rapidly growing number of interrelationships affecting humanity]."[3] This reshaping helps us find expedient solutions to current problems, but will it work over the long haul?

Inventor Allen R. Kahn and his assistant Berry C. Deer suggest that because we rely on what we already know to "select and modulate incoming information, . . . truly novel information is ignored."[4] We fail to notice some of the things we most need to understand in order to survive because we have no stored data that can help us make sense of them—or because they relate to things we would like to forget.

We are also likely to ignore what is new to us because we are uninterested in what is not familiar. When we attend a lecture or buy a book we generally select a subject we already know, and an author or speaker we know thinks as we do. As psychologists say, we like to be reinforced in our beliefs.

To meet the challenges we face, our brain developed to do specific things in circumscribed ways. Neurobiologist Michael Gazzaniga explains, "Our brains are built to process things in certain ways, and no amount of education or training can take us beyond these built-in characteristics."[5]

Unfortunately, much of the data we now need to deal with is of a nature and quantity that we are not equipped to handle. Gazzaniga writes, "Evolutionary processes have probably not conferred on the human being all possible capacities to learn all kinds of strange things. There must be phenomena in the world that totally elude us because we do not have the capacity for considering their meaning."[6]

WHAT WE DO REMEMBER

Memory is not a reconstruction of reality—it is a very limited memorandum of it. We only retain what was particularly striking to us and what our limited knowledge told us was important. Details and nuances that may turn out to be important at a later date may be omitted. Some things we gain early in life—knowledge, beliefs, and habits—become etched in our memories and can be hard to change. Recent learning and experiences, too, may be very vivid in our memory and strongly influence our thoughts and actions. This is good because the vast amount of data in our memories would confuse us if it were all equally accessible. Early learning guides us and provides us with many basic skills and lessons; recent experiences present us with data about current concerns.

This leaves much in memory that is not as easily recalled. When it is, it may not be as clear or felt to be as important as our early or recent learning. This was fine for early humans, but it is not today. For our own and society's benefit, we have to work with and objectively evaluate data we have gathered over a long time period.

When we pass the scene of an automobile accident on a highway, we drive more carefully—for a while. However, as time passes the impact of this memory fades and we gradually revert to our old driving habits. We can expect that our recognition of other possibilities, such as a nuclear disaster, will have similar fates when they are not persistently reported by the media and time moves on . . . at least until we begin to feel the effects.

SCHEMATA

As discussed earlier, when the amount of relevant information is too large and relationships too complex, our brain selects, eliminates, and simplifies it to a level that we can handle. We have a number of techniques and tools that help us do this. One of them, called *schema,* is used to organize knowledge into packages that help us make sense of things. We generalize and sort information into categories that can help us understand new experiences by relating them to what we already know.

Schemata greatly extend the power of a limited amount of knowledge. For example, when I see a well-groomed young man in a grey business suit come out of a prestigious office building, I immediately attach a number of qualities to him before I actually learn anything about him. They are my concept of what a man of this description might be like and it guides me in my dealings with him. This practice saves time, but it can give me a very wrong idea about the person. Some criminals who are aware of the importance of such impressions dress neatly in high quality, conservative clothing, thus lulling their victims into a false sense of security. Likewise, corporate public relations departments may cleverly hide the polluting activities of their companies behind images of public service. Although such images can mislead us, without schemata our minds would be so overburdened with learning every time we confronted a new situation that we could not function at all.

Our need to rely on schemata introduces problems. We can mistakenly associate perceived data with the wrong schema. Or we can call up a schema that is seriously faulty, or one that is basically accurate but differs in several important details. We then attach these errors to our new information. During the Cold War, for example, people sometimes saw ruthless dictators as friends because they opposed communism—which in itself is not evil (although many people believed it was). Additionally, many of us believe that because technology has solved many serious problems, it will always be able to do so.

People act differently based on what they know. Some information is objectively incorporated into our schemata, but sometimes prior knowledge twists data to reinforce misconceptions. There is an observation that puts this in an interesting way: "When your only tool is a hammer, all problems look like nails." Prior knowledge influences one's value systems and hence behavior. Thus, the significance given to natural surroundings is different for a Florida land developer than it is for a Seminole Indian.

Schemata focus our attention on what we already know, and thereby reduce our interest in learning. What we already know seems important and what is new to us can easily be seen as irrelevant, particularly if it does not please us or fit our biases. This inclines us to cling to opinions and beliefs that may be erroneous and dangerous. We may find it hard to accept into our schemata unappealing facts such as our own wastefulness, the possibility of nuclear war, or environmental degradation. Thus we develop schemata that give us an unbalanced or overly simplistic view of reality and our place in it.

Although stereotypes are essential in order for us to function at all, they may lead to bigotry and hostility. We have an inborn wariness of strangers. Fear can cause us to ascribe negative qualities to people we do not know. Not knowing strangers personally, we draw on our schemata and by so doing may deal with negative images rather than the unique human beings on our doorstep. Our distorted images of people become more real to us than the individuals themselves. In fact, we often close ourselves off from a reality that contradicts our prejudices—and do not see or hear what is around us. On the other hand, by avoiding the risk of prejudice, we risk being innocent about the world and so open ourselves to exploitation by others.

DEFAULTS

We believe we are conscious most of our waking hours. But are we? We spend much of our lives running on automatic pilot, adjusting course only occasionally.

To relieve ourselves of the need to constantly analyze things and make decisions, most of the time we operate by default. That is, we use predetermined ways to respond to specific situations. Having started to drive downtown, for example, we can daydream, listen to the radio, and before we know it, find ourselves in a parking garage without having given any serious thought to how we got there. While some of our defaults are inborn, many are routines we have created for ourselves. Once established, they operate with little or no conscious effort.

Occasionally, we arouse ourselves and make very conscious decisions, or look over a certain default, evaluate it, and modify it. This is not often the case. However, when we are confronted with something for which we have no default, our perceptual system arouses us and sets our mind to work planning a response. We can set and modify our values and goals and establish behavior patterns that help us attain them. We can then automatically follow the course that has been laid down, our defaults reacting to situations as they arise.

Simple default habits, such as jumping into a car and driving two miles for a pack of cigarettes, when indulged in by enough people, can have damaging consequences. This particular example increases global warming, speeds the depletion of petroleum reserves, and increases the number of automobile accidents, not to mention what it does for the number of lung cancer cases.

We can easily develop a default that externalizes causes for troubles and places blame beyond ourselves. Routinely blaming our colleagues, enemies, homosexuals, welfare, blacks, Jews, or the weather instead of ourselves for things that go wrong relieves us of accountability, guilt, and a feeling of inadequacy. This absolves us of responsibility, allowing wrongs to continue.

Inborn defaults that were suitable in a primitive society may be dangerous today. For example, the desire to hunt and kill animals can lead to their extinction. Additionally, our defaults do not respond to many of the slow changes taking place around us today. We are unable to notice events such as the depletion of the earth's ozone layer unless we are warned by specialists who have conducted studies with scientific instruments. Our ambition and our drives for comfort and security, along with peer pressure, trigger defaults that may be out of step with reality and thus dangerous for modern society. They can lead to conflict, the waste of natural resources, and destruction of the earth's protective ozone layer without our ever consciously deciding to do so.

OUR DIVIDED SELVES

The man you see on your television screen drearily talking about the national budget and the man in front of you at church leading you through an age-old ritual seem to be somber, responsible people. However, tonight with their wives they may act like wild, passionate animals. They are both, but at different times.

Although I see myself as a single, unified, coherent person, I am aware that my emotions, concerns, and interests change. Sometimes I am objective, cool, and logical. At other times, basic primitive urges take over: aesthetic feelings prevail, fear overcomes me, I want to enjoy myself, or all I want to do is sleep. At one time I seem to be one person, at other times another. I am not consistent. The connections among my various parts are weak compared with the ability of these parts to overpower other parts at different times.

Despite what many people believe, different parts of our minds may not be connected with one another functionally and those that are may communicate in disparate codes, making direct communication between them impossible.[7] This may account for our inconsistent behavior and for the difficulties we often encounter in linking related facts or ideas in our minds. Michael S. Gazzaniga writes, "A confederation of mental systems resides within us. Metaphorically, we humans are more of a sociological entity than a single unified psychological entity. We have a social brain."[8]

From a different direction, psychologist Robert Ornstein writes,

> . . . we do not have access to all of our talents at once. Our consciousness is clearly limited to only a few items at a time. This is why the system comprises an uncountable number of small minds.
>
> We wheel in and out of consciousness a certain number of these small minds to handle different situations in our lives.
>
> So, at any one time we are much more limited, much more changeable, than we might otherwise believe of ourselves.[9]

Marvin Minsky, founder of MIT's Artificial Intelligence Laboratory, sees the problem from yet a different perspective. "I'll call 'Society of the Mind' this scheme in which each mind is made of many smaller processes. These we'll call agents. Each mental agent by itself can only do some simple thing that needs no mind or thought at all. Yet when we join these agents in societies—in certain very special ways—this leads to true intelligence."[10]

Not only are our minds made up of many disparate parts, as we can observe in ourselves, but these are often in conflict with each other. For example, one part of our mind tells us that we must lose weight, while another part tells us to go ahead and eat that piece of cake. At one time the more disciplined part dominates, at another time, the self-indulgent part does. As Robert Ornstein observes, "The separate mental components have different priorities and are often at cross purposes, with each other and with our life today, but they do exist and, more soberly, 'they' are us."[11] And Marvin Minsky notes, "We all sense feelings of disunity, conflicting motives, compulsions, internal tensions, and dissension. We carry on negotiations in our head. . . . But if there is no single, central ruling self inside the mind, what makes us feel so sure that one exists? What gives that myth its force and strength?"[12] It seems that how we like to see ourselves falls short of reality.

OUR TROUBLE CONNECTING

In 1983, twenty-eight-year-old Nancy Cruzan was involved in an automobile accident damaging her brain in such a way that only the stem, which controls certain vital functions, continued to work. She was not even able to breathe or swallow. The Missouri courts insisted, against her family's wishes, that she be fed with a tube. They did not see a larger whole. They saw only her, and placed a very high value on her maimed life.

They did not consider that at the same time, 23,000 of the world's children died each day as a result of undernourishment, and many more became brain damaged for this same reason—many in the United States. How could they justify allocating funds to keep one person alive in a comatose state when hundreds of otherwise healthy children could have been helped instead? Is this failure to connect a deficiency of our minds, a way to make decision making easier, a manifestation of chauvinistic nationalism, or all of the above?

Today we must deal with interconnections in the world around us. However, as noted earlier, this is not easy, so often we don't. When we are aware of related ideas in different sections of our minds, with effort, we can often connect them. But how well can we do this? Because much of what we know about the brain comes from studying the effects of injury, the following example may shed some light on this.

People who have suffered damage to the left inferior parietal system of their brains, where much symbolic processing (such as reading and writing) takes place, may not be able to grasp the meaning of sentences although they can understand each word perfectly. This pathological condition hints that it is possible for a normal brain to have difficulty comprehending wholes composed of easily understood parts. That is, we may have trouble joining different pieces of data into the whole units that properly represent reality, even though we can see them as separate and intelligible entities. This difficulty in connecting helps account for our often unbalanced view of life and the world around us.

We especially have trouble recalling, connecting, and utilizing information that is only indirectly related to a particular situation—significant as it might be. It they were to think about it, most economists, politicians, and businesspeople would know that neither our nation's population nor its economy can grow indefinitely. Yet when these things fail to grow, such people register alarm. While they are understandably concerned about the problems that accompany a stagnant economy, they totally ignore the consequences of perpetual expansion. It does not occur to them that many current problems of our larger cities are the inevitable result of successful development.

We tend to deal with events and things as isolated, rather than parts of an integrated whole—often to our detriment. History provides many examples of this: The Aswan Dam in Egypt is an engineering marvel, but it is an ecological disaster. Among the problems it causes are the elimination of floods, which deprives farmland of its annual replenishment of rich topsoil; reduced river flow, which increases the salt content near the delta and reduces the fish population; and an increase of parasitic organisms, causing serious illness in humans. The control of malaria and other diseases in developing countries was extremely effective, whereas the resulting unplanned population explosion, which is only beginning to reveal its more serious repercussions, is devastating. Sadly, many problems we face today are the results of sincere efforts to improve conditions for individuals and society.

Urban freeways were designed to relieve traffic congestion. Those wide, traffic-light-free lanes and low taxes on petroleum encourage people to move farther from the centers of cities. The sprawl this produces requires people to drive more, which only worsens a number of problems our government is trying to reduce. Freeways have destroyed neighborhoods and led to the decline of central cities while producing more pollution and increasing our need for foreign oil (which further accelerates our trade imbalance). The greater number of auto accidents results in increased medical expenditures. Productive farmland is turned into subdivisions, and huge parking lots that need to be lighted lead to the strip-mining of beautiful Kentucky landscapes. Refrigerants leaking from air conditioners significantly contribute to the depletion of the atmospheric ozone layer. Obviously, we do not see the links between freeways and the damage they cause.

Several factors reinforce our propensity not to connect ideas. Single-minded thinking is not too arduous and often produces visible rewards. Holistic thinking can be difficult and can lead to frustration and guilt. Most of us like to avoid complex ideas and subconsciously repress unpleasant facts. We do not want to be reminded that our wasteful lifestyle will leave a depleted, polluted planet for our descendants. While we usually operate in a fractured mental condition, we are unaware of it.

In simpler times (and in some ways even now), limiting the scope of one's thinking had survival value for individuals and society. A nation is more secure when in battle its soldiers focus their attention on fighting and not on food or sex. A baby is safer when its mother's desire to protect it overpowers fear and selfishness. However, the singlemindedness that makes this possible can be devastating for our planet and society today.

Everything in the world is connected and interacts in some way. When there were few people and they had only a small impact on their environment, this fact could be ignored. However, today as we destroy the stability of the biosphere, it is highly dangerous to ignore connections. But this is what we are doing. Our fractionated minds help keep us unaware of our narrow picture of reality.

COMPREHENDING COMPLEXITY

We can only deal with a limited number of things at once, and only a few of those effectively. If we seriously involve ourselves, as ideally we should, not just with our children's education, but with starving Africans, human rights in Central America, the economy, the preservation of historic buildings, cruelty to animals, the welfare system, and nuclear waste, we burden ourselves with more than we can possibly handle and become neurotic and ineffective. However, without public understanding and concern, issues do not receive adequate support. Our inability to cope with and care about a broad range of issues at one time is a serious problem today, when public support for every one of them is needed.

As I mentioned earlier, the filtering mechanisms of our cognitive mind reduce the amount of data that passes through it into other parts of our brain. But even this may leave more than our minds can work with. We must reduce and simplify information further. Often we do this successfully and achieve positive results overall. However, with the best of intentions, we may overlook seemingly irrelevant but critical factors that ultimately can turn our efforts into disasters. For example, economists ignoring resource depletion and environmental factors produce faulty models upon which important decisions are based.

Where we can eliminate data, we must make the right choices. A number of things make this difficult. We may not know or understand enough about the problem we are dealing with or about the factors that influence it, or we may fail to analyze them clearly. Personal values and prejudices, and a preference for working with what we know can affect our choices. Correctly choosing what to ignore is not enough. We must give proper weight to what remains.

Kenneth Boulding, the late unconventional economist, described another problem we encounter in dealing with our multifaceted world—our difficulty in seeing the difference between apples and oranges:

> The human imagination can only bear a certain degree of complexity. When complexity becomes intolerable, it retreats into symbolic images. We have an intense hatred, for instance, for multidimensional value orderings. We cannot be content, for instance, by saying that John is better than Bill in mathematics but worse in history. We want to put John and Bill on a singular linear scale and say, at the end of the year, John is "better" than Bill—or vice-versa.[13]

Our failure to differentiate became particularly evident during the Reagan administration. President Reagan convinced people that government programs are wasteful and bad and that spending for them should be reduced. So programs—including assistance and education programs (in fact, nearly all programs except defense)—were cut across the board without evaluating the programs' effectiveness or the long-term results of such reductions.

IMAGINATION—OR LACK OF IT

As I will discuss in chapter 3, there are two ways we learn about the world beyond our skin. Sensation, which tells us what is happening to us, is far stronger than cognition, which tells us what is happening beyond us. Extending this point: What we see and feel ourselves is stronger than what we learn intellectually. The more imaginative we are, the better we are able to narrow this gap.

Imagination has a civilizing influence on us. It enables us to empathize with other people and even animals. It allows us to picture people in far-off lands living under different political systems in other eras. Imagination helps us to visualize different possible futures and to construct "what if" scenarios. It prompts us to extend our concern beyond ourselves.

Individuals, organizations, and governments are continually confronted with situations without precedent. They must not only react to situations for which they have no direct past experience to guide them, but they also must plan and set policy. We must deal with situations that extend far beyond our immediate surroundings and reach farther into the future than we ever had to concern ourselves with before. What we do today, whether as leaders or consumers, influences others we cannot see. Good imagination, on the part of both citizens and their leaders, is not just desirable but essential. Unfortunately, this quality is all too rare.

The strength of imagination varies greatly from individual to individual, but on the average, our imagination is weak. Whereas it was suitable for the needs of our simple past, now it is totally inadequate. My own observations and experiences with people who occupy positions of power indicate that such people are likely to have particularly weak imaginations, which causes them to be seen as "sound practical people" who can be trusted. Such leaders have an easy time ascribing poverty to laziness and environmental concerns to extremism, and can enter a war with little empathy for the victims.

Insufficient imagination can be dangerous. American industry fell behind Japan because its leaders failed to see changes in the business environment and could not picture a future much different than the present. Global warming, accompanied by probable climate change and the flooding of coastal areas, does not arouse people the way a recession or tax increase does. Problems close at hand or similar to those we have already experienced are easy to visualize, most others are not.

An inadequate imagination makes it difficult for us to understand or sympathize with people who are different from us in sexuality, social position, religious belief, race, or nationality. In dealing with other people, we have a built-in tendency to believe that they think the way we do. Many don't, which can be hard for us to accept. Misunderstandings that arise from our assumptions can be the source of many difficulties, and they make cooperation and problem solving difficult.

An additional source of much of the world's misery today is a lack of empathy on the part of well-meaning people. Virtually all Christians know the parable of the Good Samaritan, but many fail to see how it applies to them. Blind to the poverty and hardships in the world about them, they practice charity by donating a stained glass window (with their name prominently displayed below it) to their church. Actually digging in and providing food and such to the poor and less fortunate would be a far more effective gesture. Our gut feelings are often dangerously out of sync with reality.

Perhaps the most problematic consequence of our meager imaginations is this: Although we may know better, we accept our environment and lifestyle as being normal. It seems normal to drive between five and twenty miles to a shopping center and to vacation at a ski resort in Colorado.

The fact is that our planet can only support a fraction of the world's people living this way and it can continue that for only an instant in human history. The persistence of present trends, resource depletion and environmental degradation, for example, will lead to a horror beyond anything we have experienced before. Yet we lead our lives as if we can extend our present or even a higher standard of living to every member of an exploding world population. Though irrefutable evidence is before our eyes, many of us just cannot comprehend it.

When we encounter a situation, we normally rely on our past experience for guidance. Confronting something new is not easy. For example, large numbers of Americans enthusiastically welcomed our entry into World War I. It took the real experience of war to awaken them to its horrors. Similarly, governments did not take air pollution seriously until after a series of disasters, including a 1952 temperature inversion in London that caused the death of over 4,000 people. Vice President Al Gore, while still a senator, associated this difficulty in processing new information with the environmental crisis: "Perhaps because it is unprecedented, the environmental crisis seems completely beyond our understanding and outside of what we call common sense."[14] We seem to only learn from experience, but in the case of nuclear war or climate change, this is a poor way to learn.

OUR OVERRATED POWER OF REASON

We have self-flattering illusions about our ability to reason and think clearly. For example, just notice how convinced you are of your own correctness in an argument. If we were as logical and honest as most of us think we are, most arguments would be quickly settled (the correct position would be recognized right away) and all parties would depart enlightened.

Current problem solving among the majority of people seems to entail making a decision based either on what is currently fashionable or on what will reap rewards from the powers that be—although either result is likely to be faulty. Very few people will look at the problem objectively, penetrate to its root causes, and rationally solve it in a sensible way. If you push people to do this, they either won't understand what you want or will become irritated, or both.

Logical thought is not our forte. It demands greater precision than our minds can provide and more complete and precise data than is generally available to us. We would be in great trouble if we had to depend on our ability to reason.

Evolution has found a system of thinking for us that largely circumvents the need for logic. Author Jeremy Campbell explains:

> Our everyday reasoning is not governed primarily by the rules of logic or probability calculus, but depends to a surprisingly large extent on what we know, on the way our knowledge is organized in memory and on how such knowledge is evoked. Highly intelligent people, asked to solve a simple problem calling for the use of elementary logic, are likely to behave like dunderheads unless the problem is couched in such a way as to trigger networks of knowledge that are organized in logical fashion. Content is of critical importance in deciding whether we think rationally or not, whereas in logic exactly the reverse is true: Content is supremely unimportant, and form is everything.[15]

Our previous knowledge and biases largely determine which data we use in our thinking processes. We therefore base many important decisions on partial or biased information, which, when it depends on association rather than logic for decisions, can easily produce erroneous conclusions. These erroneous conclusions become new knowledge and in turn distort new data selection and thought processes. This has negative implications when dealing with reality which includes complex problems such as those involving politics, international relations, economics, and ecological systems.

When things around us seem unclear, irrational, or disorderly, we feel insecure. As I will explain in chapter 7, this creates in us a strong need to explain what is happening and why. However, the poor quality of available information combined with our limited mental capacity often makes correct explanations impossible. We rarely let this stand in our way though. We invent plausible, though often incorrect, explanations. Our existing knowledge and biases play an important role in accomplishing this. At times our need for "answers" is so great that we are driven to accept nonsense; not even our better judgment can resist it. We can even become fond of false or nonsensical data and the bizarre conclusions we derive from it. And on the other hand, as Gustave Le Bon, a French psychologist who wrote in the late nineteenth and early twentieth centuries, explains, we can equally well avoid the truth. "Evidence, if it be very plain, may be accepted by an educated person, but the convert will be quickly brought back by his unconscious self to his original conceptions."[16] Psychologist Jean Piaget observed that children tolerate contradictions for a relatively long time during their development.[17] Maturing is supposed to eliminate early contradictions. However, many people are unable to overcome the inability to recognize contradictions.

We also have great trouble balancing risks. Some people who would never get into an airplane are convinced that electromagnetic forces cause cancer. Others are afraid to walk on city streets but do not hesitate to jump into their automobile, and, without seat belts, drive along busy freeways while listening to broadcasts recounting current accidents on those very roads. Irrefutable statistics supporting the benefits of wearing seat belts do not influence them.

Most of us cannot get the concept of exponential growth into our heads. We know that the world's human population keeps doubling in increasingly shorter periods of time. Reason would tell us that because space is limited and food cannot be produced quickly enough or in large enough quantity, this will quickly lead to disaster. However, the seriousness of this situation does not hit home. After all, the population of the world has doubled many times before, and it has not caused any serious problems to me and my family or anyone around us. Experience, or perhaps lack of it, overpowers the clear case made by reason.

In his book *Inevitable Illusions*, cognitive scientist Massimo Piattelli-Palmarini describes a limitation in our thinking which he calls "tunnels."[18] Our thinking is directed into narrow channels that affect our thoughts. Just as we have visual illusions, we also have cognitive illusions. Some of these tunnels, such as "overconfidence" and the inability of untrained people to interpret statistics objectively, can have dangerous effects on personal and public decisions. Government officials may be certain that the earth will be able to produce enough food for everyone far into the future, and act accordingly. Even though we are presented with powerful facts and statistics (many reports have shown that the food supply *cannot* keep up with the population), the reality of many dangers just does not sink in.

It would seem that our inability to think logically would disturb us and that we would work hard to overcome it. For most of us, however, this is not the case. In fact, we do a poor job of using the limited abilities we have. We are unaware of how poorly we think and are quite satisfied to let it go at that.

Our minds are basically the same as those of our hunting-and-gathering ancestors. This mind has been able to develop the complex society we have today. However, some of the very qualities that helped our species evolve and survive now threaten our future.

While our brain is the cause of many of the dangers we now face, it can also help us overcome them. We will look at this later in the chapters on solutions.

3

Those Ever-Compelling Primary Drives

Walking along the edge of the Boston financial district one evening several years ago, I came upon a long line of people. They were all young, dressed in conservative business clothes, and in surprising good humor for people waiting in a line that hardly seemed to move. I followed the line around a corner to discover that it went into a bar. The room was so full of people I couldn't imagine how anyone could raise a drink to her lips, or why anyone would want to go in there in the first place.

I asked a group of nicely groomed women in the crowd, "Why are you waiting to go in there?" "To meet our friends," one curtly replied, having taken offense to my question. Around the corner, near the line's end, there was an almost identical bar, and it was nearly empty.

There was no logical explanation for the behavior of these people. They simply wanted to be "in"—part of the crowd. Their willingness to stand in line was motivated by one of those basic urges originating far in our past.

The oldest and most primitive parts of our brain, which we share with frogs and lizards, are close to the spinal column. They control the instincts, drives, and emotions that have been needed for the survival of our species. Hunger impels us to eat when we need food, the sex drive causes us to procreate, and fear initiates our defensive or offensive mechanisms. Primary drives originating here have a powerful influence over our actions, making it difficult to bring higher level functions such as reason to the fore. In the words of Bertold Brecht from his musical play *The Three Penny Opera*, "First comes the food and then the morals."

Instinctive drives that served our forefathers well are often inappropriate in the physical and social environment of the late twentieth century. Reactions once helpful for survival—and sometimes still useful in the economic environment—may now lead us to extinction. Reason is often required, but we are poorly designed to utilize it. In a complex society ethical behavior is essential. However, while evolution has provided us with feelings that cause us to treat the people we live with reasonably well, it has not given us an inherent system of ethics that can effectively counter powerful instincts.

Normally, primitive drives dominate. We are quick to start a war, but poor at ending it—in spite of the obvious costs to both sides. World War II could not be brought to a conclusion in Europe until Germany was devastated and Hitler had committed suicide. As I write, hostilities continue in Afghanistan even though an armistice was signed years ago. We waste energy by driving powerful, air-conditioned cars even though we are aware of global warming. Our basic urges cause trouble in another way: They distract us. Our sex drive, our love of competition, or our desire to be entertained can absorb us and divert us from things that desperately need our attention. We may so involve ourselves in following professional baseball that we remain totally ignorant about political candidates before an election, or about the need to solve environmental problems.

Fear, for example, protects us from many dangers, but can introduce other perils. It is a powerful force. Once aroused, it can start a chain of irrational thoughts and passions. It can begin a cycle of suspicion and hostility leading to violence. Fear of failure can stop us from doing something beneficial.

Other emotions distract us as well. Vanity clouds our perception of reality and often prevents us from recognizing and solving serious problems. Jealousy and excessive ambition warp our thinking and undermine the cooperation that is necessary for groups to function constructively.

Selfishness played a constructive role in evolution. Each member of a species and each species as a whole acted selfishly to ensure its existence. Every species has to provide for and protect itself. This is achieved in different ways. In the case of insects, it is sometimes done by each individual taking care of itself and following instincts, such as breeding and nest building. In the case of bees, it is done by individual self-sacrifice for the good of the colony. In human hunting-and-gathering societies protecting the species followed a hierarchy. First came the individuals. Then came the immediate family, and finally the hunting group itself. At times this order was to some degree reversed. Families and tribes survived to reproduce when there was a willingness to take risks for the good of the group. In very recent human history, many people have broadened their area of concern to include their nation, religion, and special interest groups. Few, however, have extended it to include the whole human race.

The selfishness that was necessary in prehistory to ensure human survival must now give way to an interspecies generosity to continue that same survival. Today we must give our attention to a wide range of issues. We must limit our numbers, protect the environment, contain dangerous wastes, and prevent biological warfare. We must not only care about all human beings, but because we so greatly influence them, we must care about the biosphere and the evolutionary process as well. Unfortunately, we have no built-in drive that causes us to do this.

WHAT WE REALLY CARE ABOUT

When we are late for a appointment, to save a few minutes, we may take chances on the road that we normally wouldn't. We might even risk our lives. The level of danger we are willing to expose ourselves to for a mere matter of promptness is out of proportion to the importance of safeguarding life.

A review of our interests reveals other bizarre facts. We devote much more time to learning about the features of a new car, early recordings of operatic tenors, or football scores than about the safety of nuclear reactors or the contamination of our water supply. We rarely hesitate to pay for war or the preparations for it, but we are stingy when it comes to funding projects that promote peace. We open our purses when it comes to funerals, headstones, and endowing human remains for perpetual care. When it comes to more forward-looking matters like our children's education and school budgets, however, we are penny-pinchers.

A visit to a magazine stand, a supermarket checkout counter, or some hours spent watching network television is embarrassingly instructive. It shows where our interests really lie. There is much to divert, thrill, and titillate. There is too little that helps us understand the world around us and to become better community and planetary citizens. Some of us are ashamed of our petty interests, but others proudly flaunt their frivolity.

We humans evolved to care about ourselves and our close associates such as family and members of our group, and had little need to concern ourselves with what lay beyond our own small field of activities. Today our interest in people extends to stories about the private lives of acquaintances, fictitious characters, celebrities, and public figures, but for many people it does not go much further. Our disinterest in expanding our knowledge seals us off from much of reality and leaves us unaware of many situations that desperately need our attention.

The Power of Sensation

We have two means for receiving information about the outside world: sensation and perception. Sensation is totally subjective and tells us what is happening at our body surface: we hurt, we enjoy, etc. It is direct and powerful. Perception is weaker and more complicated. It must build an image in our mind that describes the outside world. Because sensation is so much stronger, it dominates our attention and strongly influences what we do.[1] People caught up in personal problems are usually overwhelmed by sensation and are unlikely to be objectively concerned about broad issues.

In the past we did not need to be aware of processes such the carbon cycle, which is essential for all life on the planet. The cycle was self-regulating and little affected by human activity. Today, however, we are dangerously changing its balance by burning fossil fuels and reducing the earth's vegetation. It has taken us a long time to learn that we are now reducing the number of species on the planet at a dangerous rate. Our response is halfhearted because we do not yet feel the consequences. The breadth of our concern is inadequate for the serious threats we now face.

AVOIDANCE BY RATIONALIZATION AND SELF-DECEPTION

Every day we are bombarded by uncertainties and unclear or disagreeable situations. In response, we use techniques of self-deception and rationalization to relieve anxiety, ease our conscience, and help us avoid responsibility. Who really wants to learn that the coolant in their air conditioner will someday help deplete the ozone layer? We do not like to relate activities such as vacationing in Cancun to the global warming caused by the energy it takes to get us there, nor to think of the malnourished children living several blocks away who might eat decently for a year for what we pay for one night at our hotel.

When we find ourselves in an awkward situation or when we are questioned about our beliefs and prejudices, we feel threatened and our defenses go up. One way to avoid discomfort is to exclude disturbing information from our world models. Psychologist Daniel Goleman writes, "There is an almost gravitational pull toward putting out of mind unpleasant facts, and our collective ability to face painful facts is no greater than our personal one. We tune out, we turn away, we avoid. Finally we forget, and forget we have forgotten."[2]

An experiment conducted by Ervin Staub and his students in the early 1970s buttresses this point: "We observed as passersby saw a person collapsing on the street. Some people, after a single but unmistakable glance, turned their heads and moved on without looking again, as if to avoid any further processing of the event they had witnessed. A few of them turned away when they reached the first corner, apparently in order to escape."[3] We are all at times drawn to act this way in a world filled with more problems than we care, or wish, to face.

Moving from an inner city problem neighborhood to a suburb to which low-income people cannot follow reduces our exposure to crime and the burden of financially supporting the less fortunate. Gradually our awareness of crime, poverty, and other problems dims and may even disappear. We can more easily pursue our own interest without worrying about others or feeling guilty.

If we can find one scientist out of hundreds who says the earth can feed many times more people than live on it today, we want to believe him so that we do not have to alter pleasant habits. The idea that "trickle-down economics" will bring prosperity to everyone is tremendously appealing. We do good by pursuing our own interests rather than by sacrifice. Soothing books and lectures with titles like "Everything Is Getting Better—I'll Bet on It!" are popular with many businesspeople and politicians. Books, articles, and lectures like this, sometimes funded by conservative foundations, help put uncomfortable thoughts to rest. As a result, many people have come to believe that environmental problems have been blown out of proportion. Consequently it has become harder to pass new protective laws and uphold old ones.

Religious beliefs can also relieve us of responsibility. Some Christian groups, by choosing to remember the Bible passage about populating and subduing the earth and ignoring ones about responsible stewardship, have fought hard to halt population and birth control programs. Also, where convenient, we readily accept partial truths as whole truths, ignoring their incompleteness or conflicting data.

Our inner need for what we believe to be order, which will be discussed in chapter 7, can lead us to see order where none exists, or where we do not have the energy and commitment to pursue it. Many other devices aid us in deceiving ourselves and others, including rousing slogans and jargon, diverting our attention, relying on faulty authority, and even defaming people or ideas. Euphemisms such as "atoms for peace," "the peacemaker," and "Yankee ingenuity" help us hide the truth from ourselves. Statements like "people are poor because they are lazy," enable us to be selfish without feeling guilty. Group self-deception goes a long way toward relieving the conscience.

One way in which this self-deception is perpetuated is through the use of labels. For example, today environmentalists are sometimes referred to as "extremists." When used with skill, labels can crush doubts and facilitate desired mind-sets. When they help us feel comfortable, we eagerly accept such terms with little thought. Fortunately, oversimplification occasionally goes too far and backfires, as when former President Reagan informed us, in his authoritative manner, that, "Approximately 80 percent of our air pollution stems from hydrocarbons released by vegetation."[4]

We are inclined to be hostile to the messenger who brings bad news. This also allows us to continue deceiving ourselves. Discrediting, and sometimes even persecuting, the messenger of bad news seems to invalidate the message and helps put disquieting information out of mind. The Bible is full of stories of animosity toward prophets who brought warnings from God. In *An Enemy of the People,* Henrik Ibsen's character Dr. Thomas Stockmann is persecuted because he reveals the disturbing truth about his town's famous spring water—it is contaminated.

Sometimes we simply take an easy way out of a difficult situation by diverting our attention and seeking pleasure. This route can become a habit, or as Al Gore points out, an addiction.[5]

When it does not much matter whether we do something today or tomorrow and the task requires sacrifice, the temptation to do it tomorrow is compelling. We trick ourselves into believing that we will certainly get to it, but we rarely do. Unfortunately, decisions that shape our long-term future can always be put off for one more day, whereas those affecting short-term problems often may not.

Even when we do our best to confront reality and act in a responsible way, without procrastinating, it may not be easy. While we are aware of our conscious thoughts and have some influence over them, we understand little of and have little power over our subconscious thoughts, even though they exercise considerable control over our actions. In a conflict between feelings and reason, it is the subconscious that has the advantage. Psychologist Daniel Goleman writes, "It is a simple step for the unconscious mind to act as a trickster, submitting to awareness a biased array of facts intended to persuade the aware part of mind to go along with a given course of action. The unconscious, in other words, can manipulate the conscious mind like a puppeteer his marionettes."[6]

When we deceive ourselves and others, many important issues and problems are ignored, shortchanged, or dealt with improperly. The fact that we so easily resort to self-deception, and are so good at it, implies that it must have served a useful purpose in our evolution. Unfortunately, today it can be fatal.

WE RESIST WHAT IS UNFAMILIAR TO US

When I was in high school, I spent a summer working on a farm. One of my jobs was to clean the cow barn after milking—no small task considering the several hundred pounds of what I had to remove. One day, at the farmer's request, I spread crushed limestone over the cleaned floor for hygienic reasons. It was beautiful, a real improvement, I thought. That evening when the cows came in from the pasture for milking, they took one look into the barn and refused to enter. They were terrified. They did not like what I had done to their safe, familiar floor. We had to return their floor to its original state as quickly as possible so they could be milked that evening.

Like the cows, some people get upset if their habitats are altered, if a piece of furniture in their favorite room has been moved, for example. They will not rest until the sofa or chair has been restored to its "proper" place. "Experts" also resist change, even when it makes good sense. Armor-clad steam vessels used in the American Civil War proved superior to plain wooden ships with sails. Nevertheless, shortly after the war, the Navy declared that these modern ships were unsuitable and should be decommissioned.[7]

Good Reason to Resist

There is a good reason that evolution has made higher animals, humans included, fearful and ready to flee from what they perceive to be different. What is different can be dangerous, and fear and suspicion are important survival mechanisms that warn us of possible dangers so that we can react appropriately.

Most of us want, and need, to live in an environment that makes sense to us, where we know what to expect and how to act. Stability is essential. We know how to deal with things as they are. New ideas can be threatening if they challenge our area of expertise, undermine our knowledge, and render us unskilled novices in areas where perhaps we once excelled. We have much invested in the status quo, and the arbitrary introduction of the unknown can easily lead to chaos. Normally, except for small changes, we want to keep things as they are.

What we fear comes in a variety of shapes and sizes. We fear change, new ideas, and strangers, for example. Any of these might pose a threat to us. Our aversion to what is different often stems less from the reality of today than from the possible threat alien things posed to our species during its development.

In hunting-and-gathering societies, it was necessary to be wary of strangers. They could be hostile, a threat to a limited food supply, looking for tribal secrets, or perhaps planning to run off with the leader's daughter. As with other territorial animals, these fears helped scatter people at low densities over large areas of land so that their surroundings could support them. Now, however, we rely on distant strangers for food, goods, entertainment, and technical advancement, and people with varied skills and talents are needed to keep our own society running. We are all only parts of a complex system that every one of us is dependent on.

We've all seen animals and people whose fear or suspicion prevents them from taking advantage of worthwhile opportunities, and sometimes even from avoiding danger. They become too scared to act. Today human society and our planet are threatened with numerous serious problems—the rapid extinction of many living species, for example. To correct these problems we must alter the ways we do many things. To avoid terrible changes around us we must change, but our natural inclinations oppose it, often going so far as to reject new ideas or needed change without even bothering to investigate the reasons for them. Finding the right balance is difficult, but society must learn to change in order to survive. I will discuss how society reacts to change in more detail in chapter 5.

JOINING THE CROWD

We like to think of ourselves as free spirits, as having our own thoughts grounded in clear observation and logic. We think that our values, beliefs, and choices are our own, but often they are really what others expect them to be. The famous psychoanalyst Eric Fromm writes, "As a matter of fact, in watching the phenomenon of human decisions, one is struck by the extent to which people are mistaken in taking as 'their' decision what in effect is submission to convention, duty, or simple pressure. It almost seems that an 'original' decision is a comparatively rare phenomenon in a society which supposedly makes individual decision the cornerstone of its existence."[8]

We have an instinctive drive to adopt the ideas, values, and ambitions of those around us. For example, most people accept the political orientation and social attitudes of those with whom they associate. Psychologist David Ricks gives another example of our follow-the-leader mentality. He points out that besides "bull" and "bear" investors, there are "sheep" struggling to see what others are doing and blindly following them. History is full of idiocies instigated by people wanting to be accepted by a particular group. The Chinese elite once thought it fashionable to bind women's feet so they would always appear dainty. The corsets of the late nineteenth century were designed to give women an artificially induced narrow waist. And a desperate nation followed an unhandsome Austrian (Hitler) into spiritual and physical ruin.

It Pays to Fit In

Belonging provides us with a feeling of security and pride in the power and assumed achievements of a group. It frees us from the effort of making decisions, and from much personal responsibility. Acquiescing to others opens the doors of opportunity in business, the professions, politics, and society.

Evolution also favors a high level of conformity. Conformity helps individuals think and procreate, and groups of conforming individuals have better chances of survival than groups of nonconformists who pursue their own interests. We are inclined to feel that when many, or at least those we trust and admire, believe something, it must be true. As Fromm pointed out, "The average person who goes to a museum and looks at a picture by a famous painter, say Rembrandt, judges it to be a beautiful and impressive picture. If we analyze his judgment, we find that he does not have any particular inner response to the picture but thinks it is beautiful because he knows that he is supposed to think it is beautiful."[9]

Our ambitions and relationships push us to conform so that we do not stand out and jeopardize our position in society. While we want to be unique, we want to be only slightly so. Most of us tend to avoid discussing our deeper thoughts with others lest we give them the impression we are odd or different or offend them. The highly original, the nonconformist, the moralist, and the "whistle blower" incur the penalty of being ignored, ostracized, or even attacked. Like chameleons change their color, we often change our personality, our character, and even our appearance in order to fit in.

The Pursuit of Status

Being accepted is often not enough. Many of us need to be admired by a chosen set of peers. They may be the rich, the famous, the socially prominent, scientists, a street gang, or the collectors of baseball cards. There are various ways to gain esteem. They include money, power, status, possessing a large stamp collection, being top weightlifter at the local gym, or downing a record number of beers in fifteen minutes at Bob's Bar.

Many people own and consume not to satisfy themselves but to impress others. Small, inexpensive cars can transport people just as well as large, expensive ones. Yet some people who complain bitterly about paying more taxes to improve schools easily spend up to five times as much as they need for a car often just to impress others. Large lawns in front of pretentious homes in exclusive suburbs have few functions beyond establishing the occupants' status and keeping power mowers busy. In another part of town, adolescents rob or kill other adolescents to obtain a jacket or pair of shoes that will impress their friends.

The Kwakiutl Indians of Vancouver Island established status by making extravagant gifts of property which had to be accepted and reciprocated even more lavishly by the receiver at a later date. Not to do so was to accept defeat and a lower status. Another means of raising one's own status and subduing rivals was to destroy personal possessions, including one's own home and boats. Your rival was then expected to outdo this.[10] This rightly seems absurd to us. Our status-seeking activities are of course more reasonable—some of us merely drive ourselves into bankruptcy in order to impress others.

What gives us satisfaction is often not how much we have or how important we are, but where we stand in relation to others. This goads us to endless efforts to increase our wealth, status, or whatever. Admiration of such achievement gives us the response we crave, and encourages this behavior.

The Consequences of Fitting In

Peer pressure can sow seeds of fear and hate. Watch children join together to pick on some unfortunate—or grownups ganging up on people of different ethnicity, sexuality, religion, or whatever. In times of war or intense patriotism, writers and journalists dehumanize the enemy, and soldiers commit atrocities so that they are not considered cowards by their comrades.

Conformity and our pursuit of status can prevent us from solving problems sensibly and logically. They further conspicuous consumption and encourage us to do things with little thought about the purpose or the consequences. In order to maintain a certain lifestyle, we may overwork ourselves, neglect our children, become neurotic, and spend little time pursuing our true values. We may become committed nationalists, conservatives, radicals, religious believers, or bigots in order to be accepted by a chosen peer group. Because we are this way, unless our peers are concerned, it is hard to arouse our own concern about humanity's future or the inequitable distribution of the world's wealth—and our peers won't be concerned unless we are.

This thoughtless groupthink, where everyone thinks along the lines of everyone else, confines us to conventional wisdom and prevents us from seeing reality and thinking critically. Until very recently, a general lack of concern about environmental problems assured us that there really were no problems at all. How could they be when "everybody" did not take them seriously? We somehow cannot believe that most people, especially our leaders, can be wrong and that life will not continue much as it has.

While conformity pays off for individuals and maintains order in society, it does not come without cost, which can range from our personal freedom and integrity to damage to others and to our planet.

FASCINATION WITH COMPETITION

Some years ago I attended a weekend workshop run by a sociologist. At the start we were divided into groups and told to organize ourselves, and that later we would interact with the other organized groups. With much ado and jockeying for power, the members of my group elected officers and created positions for those left over. I wanted to discuss the goals of our organization, but in the scramble this aroused no interest whatsoever. Near the end of the weekend, we were told that we would soon start interacting with the other groups. Immediately the discussion turned to how we could get the better of them. Our workshop leader had never suggested that conflict or competition was a goal for these interactions, this behavior just sprang up from deep within ourselves. We were not interested in the purpose of the interaction, we just wanted to prevail over the other groups.[11]

We are usually more interested in winning than in doing our best or gaining understanding. Arguments often end up as shouting matches, each side trying to convince the other, with neither party interested in gaining a better comprehension of reality. Why do we persist in this behavior which only harms us in the end? We really are not all that interested in truth; we like to win; we love our opinions and convincing others of their validity; and where self-interest appears to be at stake, we dislike the truth. Our reticence to accept reality—often knowingly—makes it very hard for us to reach a constructive consensus. However, facts useful for beating the other guy, such as knowing how to improve your tennis serve or to build chemical weapons, are not ignored or so easily forgotten.

During our evolution people who successfully competed for food, shelter, mates, and other advantages produced more descendants than others. As Charles Darwin explained, it was "survival of the fittest."

Competitiveness was thus strengthened and became an important feature of our species. We reinforce this quality today by further rewarding with money, respect, and fame those who compete successfully. Famous athletes, war heroes, business tycoons, and even dictators are respected. Many of us value success more than personal qualities such as kindness, honesty, knowledge, and wisdom. Consequently, our society is highly competitive.

The prizes currently sought by competitive individuals include power, goods, fame, and sex. Those who don't participate directly become spectators. By watching the competition, they participate vicariously. Sports, politics, business, beauty, and baking have all become matches to be played and to be watched.

Many politicians are more interested in power itself, and the prestige that goes with it, than in the daily problems of governing. Many businesspeople enjoy the competitive process more than the substance of their business, be it donuts or tires. Some corporate leaders push their organizations toward perpetual expansion more for their inner need to win than to increase profits. Don Juans relish the conquest more than the object of it. During a political campaign the media satisfy our desire for show and contest by giving us play-by-play reports on the tactics and on who's ahead, but tell us little about the issues.

Competition Can Be Harmful

Competition served a useful purpose in our evolutionary development and it is still an important motivator today. But it has negative aspects as well.

People good at attaining power often achieve it through their single-mindedness. They gain positions that would be better held by individuals who understand issues and are sincerely concerned about them. These powerful, competitive people make decisions that determine the relationships between nations—whether they be cooperative, competitive, or hostile. They determine whether an environmental feature is to be protected or exploited by their political supporters. They influence the distribution of the world's wealth. Unless checked, the competitiveness of these people can overpower cooperation and drown out concern for constructively addressing pressing problems.

OUR ATTRACTION TO VIOLENCE

As a species we can be proud of our achievements in science, technology, medicine, commerce, and art, and also the many acts of kindness and generosity we have accomplished. Yet we have a terrible dark side to our nature. We are capable of appalling acts of cruelty toward others. When I was a child I thought such cruelties were something of the past and that we had progressed beyond this. I now know better. The German holocaust; the Chinese Cultural Revolution; death squads and torture in El Salvador; mass extermination in Cambodia, Bosnia, and Rwanda; and brutalities committed by Americans in Vietnam reveal what average human beings are capable of doing today.

It is not only greed, ambition, hate, fear, or lust for power that drive people to start wars and commit atrocities. There are loftier motives as well: improving conditions of a suppressed people, righting past wrongs, for the motherland, for God, or even to prevent a war. Even when motives are not worthy, though, we delude ourselves into thinking they are.

I suspect that the people who lead nations into wars are highly competitive, power-hungry, and egotistical, and have either little imagination or no compassion for those who will suffer. They may even be paranoid. But what about those followers who make the war possible and who carry out the atrocities?

The Dangerous Man Who Lives Next Door

In every society, there are obvious criminals, sadists, and people who are fascinated by violence. But when hostilities erupt, large numbers of people willing to participate in barbarity appear. Where do they come from?

For years I have wondered who around me, in this relatively peaceful society in which I live, could be perpetrators of cruelty under a tyranny. Such individuals must be here on every side—today upright, respectable citizens, but under different circumstances, monsters. Who are they? Are they among my acquaintances? Am I one of them? A number of scholars and researchers have looked into this question. Andrew Bard Schmookler writes, "The monsters among us provide a valuable insight into the human condition. The inner demons that make a Hitler . . . are to be found in the rest of us as well. . . . What we may keep hidden and under control, in a Hitler or a Stalin showed its naked face to the world."[12]

Consider the acts committed by members of a profession dedicated to the preservation of life in one of the most advanced nations on earth, i.e., German doctors selecting Jewish prisoners to send to the gas chambers at Nazi concentration camps. Psycho-historian Robert Jay Lifton, professor at the John Jay College of Criminal Justice and founder of the Center for the Study of Violence and Human Survival, made a thorough study of them. He describes these doctors as ordinary, "Neither brilliant nor stupid, neither inherently evil nor particularly ethically sensitive, they were by no means demonic figures—sadistic, fanatic."[13]

Gustave Le Bon observed a different people at a different time in history. He noted, "all mental constitutions contain possibilities of character which may be manifested in consequence of a sudden change of environment. This explains how it was that among the most savage members of the French Convention were to be found inoffensive citizens who, under ordinary circumstances, would have been peaceful notaries or virtuous magistrates."[14]

Some Experiments with Average People

A famous experiment conducted by Philip Zimbardo and his colleagues at Stanford University supports these observations.[15] Twenty-four normal, healthy, male college students were randomly divided into groups of "prisoners" and "guards" for an experiment that was to last two weeks. In order to provide a realistic environment, appropriate cells and uniforms were provided in the basement of the psychology building. Within a short time the "guards" appeared to derive pleasure from insulting, threatening, and dehumanizing the "prisoners" who became depressed, anxious, passive, and self-deprecating. These feelings became so intense that the experiment was terminated after only six days. The researchers felt a need to keep in contact with the subjects for a year to make sure that the negative effects of the experiment no longer persisted. Most distressing to the researchers was the facility with which "normal" young men could adopt sadistic behavior.

Commenting on this experiment, sociologist Zygmunt Bauman of Leeds University wrote, "The most poignant point, it seems, is the easiness with which most people slip into the role requiring cruelty or at least moral blindness—if only the role has been duly fortified and legitimized by superior authority."[16] For Bauman, the exceptions are rare individuals who are able to resist authority and assert moral independence.

Vilmos Csányi, professor of ethology (animal behavior) and behavioral genetics at the Lorand Eotvos University of Budapest, offers an explanation: Individuals with a readiness to adapt to authority produced more offspring than those who resisted those in power, giving them an evolutionary advantage. This trait is unknown among animals, but it plays an important role in every field of culture.[17]

Another famous experiment conducted by social psychologist Stanley Milgram at Yale University in 1962 demonstrates how normal people readily resort to brutality when ordered to do so. Forty male subjects, twenty to fifty years of age, participated. One by one they were brought into the laboratory and introduced to the administrator of the session and to an accomplice, whom they were told was also a subject of the experiment. They were informed that the purpose of the experiment was to find out what influence punishment had on learning. They then drew lots with the accomplice to see which one would be the "teacher" and which the "pupil." The lots were rigged so that the true subject was always the teacher and the accomplice always the pupil.

The teacher and pupil were taken to another room where the pupil was strapped in a chair, and an electrode attached to his wrist. The teacher was taken back to the original room and seated before a series of switches labeled from "slight shock (15–60 volts)," to "XXX (435–450 volts)."* He was told to give larger shocks to the pupil in the other room each time the latter gave a wrong answer to a question. As the session progressed and the teacher supposedly reached 300 volts (purely a sham), the pupil, who was part of Milgram's team, pounded on the wall between the rooms giving the "teacher" the impression that he was in great pain.

*The 110 volts in an ordinary residential outlet can kill you under certain circumstances.

At this point the teacher, feeling uncomfortable, usually turned to the administrator for advice. The teacher was told that the experiment must continue—if the pupil did not respond to a question after five to ten seconds, the next level of shock was to be administered. The pupil pounded on the wall even harder after 315 volts, but after that gave no answer nor made any sound. No teacher refused to stop before supposedly administering 300 volts and 65 percent went all the way up to 450 volts.

Milgram felt that a number of factors explain the unflinching obedience of the teachers to the administrators, who instructed them to keep administering shocks. Similar to Csányi, he believes evolution has favored people who can adapt to organized social activity and hierarchical situations. Socialization and society's system of punishments and rewards reinforce feelings of obligation and obedience. And we do not feel responsible when we act as an agent of a "legitimate authority."[18] This obedience helps ruthless tyrants gain and keep power, and provides them with people willing to do whatever they ask without questioning.

Desensitization and Separation

Christopher R. Browning, professor of history at Pacific Lutheran University in Tacoma, Washington, reports another major reason we are drawn into brutality. He studied a battalion of middle-aged, lower-middle-class family men, reserve policemen from Hamburg, Germany, who had recently been drafted and brought to Poland. Few were racial fanatics or Nazis. On July 13, 1942, in the Polish village Józefów, they rounded up the Jews, selected several hundred as "work Jews" and shot the rest. They were not forced to do this and were given the chance to refuse. Twelve of the 500 did, the rest participated to varying degrees. During the sixteen months following this massacre, they participated in the shooting of 38,000 Jews and the deportation of 45,000. After conducting many interviews, Browning concluded that the reason so few refused to participate had to do with career ambition and peer pressure. They did not want to lose face in front of their comrades.[19]

The drive to impress peers or the opposite sex can lead people (men in particular) to commit acts of violence. This trait, sometimes purposeful in the past, often causes much damage today. Not all men are dangerous in this respect, however, highly competitive people, who may also have a tendency to be violent, often push themselves into positions where they can easily harm others. Many less ambitious men, and some women, eagerly support the violent activities of others.

Lifton's study of concentration camp doctors offers explanations he calls "doubling" and "numbing" as to why normal individuals participate in cruel behavior. "Doubling" divides "the self into two functioning wholes so that a part of the self acts as a whole self."[20] One part would commit atrocities, the other would remain a loving husband and father. Perhaps the separations between different parts of our brain and the difficulties these parts have communicating with each other (as discussed in chapter 2) explain this. "Numbing," on the other hand is a "diminished capacity or inclination to feel. . . . It is probably impossible to kill another human being without numbing oneself toward that victim."[21]

Doubling and numbing play a part in many of our daily activities. We may go to church every Sunday and be entirely sincere in our intentions to be a better person, but come Monday, we are our old selves again. We double into Sunday and weekday persons, as is convenient, and grow numb to our switching. These processes also enable us to carry on as usual even though we might be aware of the damage we are doing to our environment and of the dangers of an all out war.

Lessening of Moral Concern

In addition to, or perhaps in conjunction with, numbing, a reduction of moral concern has also occurred. Ervin Staub explains why this may have happened: "Knowledge about the Holocaust and other mass killings and about torture and terrorism has a cumulative impact. Such violence represents worldwide steps along a continuum of destruction. Furthermore, the threat of nuclear destruction may diminish the seeming magnitude of 'lesser' violence. Such interconnected change can lead to a worldwide lessening of moral concern and an increase in the ease of killing."[22]

Today the media bombard us with acts of brutality from around the world and our television programs are filled with violence. Much of our time is spent absorbing this. It does not make us more peaceful.

VIOLENCE BY REMOTENESS

In our modern world of technology, too, the perpetrator is often separated from his victim, who is invisible to him. Actions—and their repercussions—become increasingly more remote from the actors. The perpetrator functions like a bureaucrat carrying out the duties of his job. In 1941 George Orwell wrote in *England Your England,*

> As I write, highly civilized human beings are flying overhead trying to kill me. They do not feel any enmity to me as an individual, nor I against them. They are only "doing their duty," as the saying goes. Most of them, I have no doubt, are law-biding, kind-hearted men who would never dream of committing murder in private life. On the other hand, if one of them succeeds in blowing me to pieces with a well-placed bomb, he will never sleep the worse for it. He is serving his country, which has the power to absolve him from evil.[23]

VIOLENCE AT A CRITICAL POINT

Our inborn bent toward violence has plagued individuals and societies for millennia, but until now, it has not threatened the existence of our species.

Conflict among preagricultural humans was not as dangerous as it is today. Population was low and hunting groups lived far apart, only aware of others who lived near them. Weapons were insubstantial, and when fighting took place adversaries generally confronted each other directly, one on one.

Today, hostilities exist between nations that are separated by continents. Religious, political, and economic beliefs often intensify any differences or conflicts that do exist, and some of these nations have weapons capable of destroying most of the life on earth. Their leaders, however, may be emotionally, intellectually, and ethically ill-equipped to handle the responsibility these armaments bring.

The human mind has not evolved sufficiently to handle this situation safely, nor have modern states made adequate use of whatever means are currently available to help maintain peace. Instead we have used our brains to develop incredibly destructive weapons and techniques for manipulating people's thoughts. These are particularly dangerous in the hands of totalitarian governments or fanatical individuals who thrive on conflict.

The obstacles to rational behavior are real, but our situation is not without hope. It is by acting on motives beyond pure instinct that we differentiate ourselves from other animals and become truly human. We often achieve this, but we can do better.

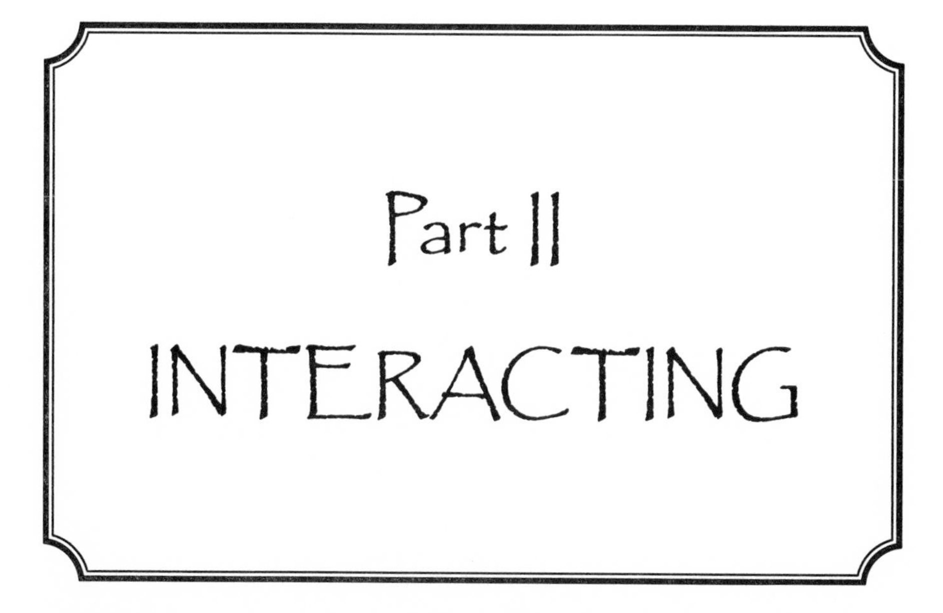

Part II
INTERACTING

We humans cannot subsist as individuals or as a society without social contact. We not only need such contact, we strongly desire it. Yet, our transactions, beliefs, and ethical systems often lead to discord, hostility, advantage seeking, and sometimes even brutality and murder.

4

When We Come Together

Throughout the history of the universe there has been a recurring movement from the simple to the complex—from electrons to the United Nations. Cooperation helped single cells join together to form multicellular organisms. Algae and fungi, for example, two distinct and separate species, join together for mutual benefit in communities we call lichens. Bees performing different tasks live together in communities that function as coherent wholes. We humans also profit by joining together to form families, communities, and nations.

As human social evolution advanced, societies needed more expertise than any single person could develop. Evolution found the solution in specialization. Individuals developed skills in specific directions and learned to rely on other people for skills they themselves did not possess. This made it possible for families and societies to have more capabilities than any single individual could possibly have. In families, different tasks were performed by men, women, the young, and the elderly. In groups, some became chiefs, priests, healers, warriors, or gatherers. Specialization is not unique to humanity, though; it is apparent in ant and termite colonies and between the sexes in many species.

As we all know, most people are stronger in some ways than they are in others. Friedrich Schiller, William James, and Carl Gustav Jung each wrote about this.[1] In his book *Dichotomies of the Mind*, systems scientist Walter Lowen develops Jung's ideas further.[2] He describes our tendency to prefer one out of each of a series of opposing pairs of ways for dealing with the world. For example, we tend to prefer either working with people (a nurse) or with things (an engineer). We are inclined to either think concretely (businessperson) or abstractly (mathematician). We like either to deal with a number of things at one time (administrator) or concentrate on a single task until it is completed and then start another (scientist).

These, and other possible preferences out of pairs of opposites, when combined produce sixteen distinct personality types. Varying strengths in one direction of each of the opposing pairs, and differences in levels of intelligence, education, and interests add variety within the sixteen clear types. This subject is complicated and need not be explored further here, but what is important for us to understand is that each person tends to have specific ways of dealing with the world that provide strengths in one direction paired with weaknesses in the opposite. Thus different people see and approach problems in different ways.

A person adept at working with things is likely to be less skilled at handling people. An individual who is oriented to the present, likes to work on a number of tasks simultaneously, and who can handle crises easily, may find it difficult to concentrate on an abstract problem over an extended period of time or think about the distant future. Some people are inclined to solve problems by intuition, others rely on reason or are guided by emotion.

Society benefits by being composed of individuals with different talents, interests, educations, and levels of intelligence. It thus has the capacity to perform a wide variety of necessary tasks. However, we cannot fully realize the advantages of this unless we value the judgments and abilities of others in areas where we are weak.

Businesses, for example, make use of people with different personality types. Some individuals are needed to carry out routines, others to manage, still others to promote or sell, to make strategies, to keep books, to mediate, and to solve legal problems. Like any organization, a business functions best where the strengths of each individual are recognized and utilized. However, often it does not work this way.

We tend to be convinced that our way of thinking and doing things is better than that of others. Because we think so differently, we often have difficulty communicating with one another. A person who thinks rationally may have trouble communicating his thoughts to someone who bases her beliefs on feeling. No amount of logic can change the latter's opinions.

SOCIETY UNCONSCIOUS AND DISORGANIZED

"With more means at its disposal, more knowledge, more techniques than ever, it turns out that the world today goes the same way as the worst of worlds that have been; it simply drifts."[3] One wishes that we could say we have progressed since Ortega y Gasset wrote those words in 1930, but if we have, it is hardly noticeable.

A society has access to far more talent, brain power, and information than any of its individual members do. Therefore it would seem that society should behave at least as sensibly as its more intelligent, sober members are able to do alone. We know that it does not, though.

Civilization has produced great art, science, and inventions, and it has organized industry and trade with great success. Through cooperation, many of its achievements far surpass those of its most gifted individual members. As a whole however, its record is very different. It falls far short of its potential. In fact, human society often behaves badly and injures itself. Sometimes commendable individual achievements are so distorted and misused that they do more harm than good. Why don't we do better when the challenges we face are so great and the stakes so high?

Individuals have the ability to act consciously and often do. Although much of the time they react to situations according to pre-established routines and prescribed procedures, organizations sometimes act consciously too, such as when they revamp a routine or examine a new situation and decide on a course of action.

Human society, however, does not act consciously. Its activity is an amalgam of many uncoordinated individual acts of personal self-interest. Today our destiny is directed by cumulative pursuit of pleasure, goods, power, security, and our surrender to many primitive drives with little rational restraint or questioning. With few exceptions, the enactment of some international laws and limits imposed on the manufacture of fluorocarbons, for example, civilization almost never collectively analyzes situations, plans, nor considers its future in a holistic way. There are a number of reasons for this.

When people come together, selfishness, greed, jealousy, and conflicting goals and values often work against cooperation. This makes it difficult to take objective, altruistic action among or between groups of people. Individuals are pleased when society meets their own needs, but are generally disinterested when it benefits the whole.

We are very aware of our personal physical limits and weaknesses, and at least occasionally admit to the limits and quirks of our minds. We rarely think, however, about the inherent limitations and idiosyncrasies of organizations and social systems. We are used to them and live with them. Without an understanding of these constraints, civilization acts unconsciously and unpredictably. In the past there was no need for individuals to think about humanity as a whole. Humans did not have the power to destroy themselves, and natural laws were all that was needed to maintain ecological balance. Today, an unconscious society bursting out of stabilizing controls is a dangerous society.

POOR COMMUNICATION AND COOPERATION

As civilization becomes more complex, understanding and cooperation become ever more difficult. Some of our most serious problems lie in poor communication. People who live or work together may think in very different ways and communicate only superficially. We shy away from exchanging serious thoughts, so we learn little about how other people really think. I was taught that one does not talk about religion, politics, or money in polite company. What an effective way to keep conversation dull, shallow, and in a sense useless!

We are less interested in exchanging ideas than in expounding banalities. When we do discuss serious subjects we fling opinions back and forth without listening to what others say. We do make exceptions and may even listen—but usually only when ideas come from like-minded peers or famous people, or when ideas are simple, can produce a profit, or are repeated many times.

Groups holding opposing viewpoints tend to avoid communication with each other and show little interest in trying to understand each other's points of view. Issues heat up and reason vanishes with the wind people generate by pontificating. Interchange, modification, accommodation, and intellectual growth become impossible. Most political conservatives and liberals, Right-to-Lifers and Pro-Choicers, Creationists and Evolutionists do not discuss their positions with each other. Instead, they fire slogans! Stubbornness and our love for the battle itself overshadow any interest we may have in discovering truth. Irrationality wins, and we all are the losers.

The abilities and knowledge we need to reestablish balance with nature and to solve many other serious problems are not concentrated in any single individual, but are spread out among us. However, we do a shoddy job of seeking out, combining, and utilizing our know-how.

For various reasons, we do not cooperate with each other on the level that is now essential not only to solve a huge number of intricate problems, but to survive. Every one of us is pulled in two directions: We want and need to be our assertive individual selves, yet we want and need to be part of something larger. Disregarding cooperation with others can simplify problem solving and bring easy success to ourselves, but only at the expense of order, coherence, and society's effectiveness. Conversely, ignoring our own integrity, reason, and needs in order to be part of a group can provide security and comfort, but it deprives society of honest input, integrity, and talents that should be available. Dictatorships thrive on such behavior.

In politics there is a tendency for people or factions to squabble and separate themselves, as was the case with the Soviets, Yugoslavians, and Somalians. Ideologies erect barriers between people, creating dangerous hostilities, and curtailing objective problem solving. For example, both communism and capitalism contain good ideas. However, in reality they also have features that create conflicts between each other and frictions among people who function within them. Sometimes nations so weaken themselves that they open themselves to outside exploitation, as was the case with the ancient Egyptians and Romans. These conflicts, often created by ambitious individuals to advance themselves, distract us from solving the major problems that threaten us all.

Destructive human conflict is widespread in the world today. We see our greatest enemies as being each other rather than the environmental dangers that are closing in on all of us. Nor do we seem to notice that the greatest obstacles to creating a better, more secure life for ourselves lie within ourselves, or how much could be accomplished if we worked together. Instead, society is composed of individuals and groups that often work at cross-purposes. Rigid building regulations make it so expensive to rehabilitate buildings that landlords cannot afford to upgrade them for low-income tenants. Overdemanding lighting codes that benefit power companies require unnecessary amounts of electricity which counteract governments' own energy conservation programs. As individual parts gain their own selfish ends, society pays. But the damage goes deeper. Such activities demoralize people, making self-sacrifice seem foolish, and set bad examples for others to follow.

UNACCOUNTABILITY

Our successes in medicine and technology have eliminated the restraints that once maintained balance in our ecological niche. Since this began, our niche has been expanding out of control at an exponential rate. There are only two ways that this expansion will in time be stopped. First, life will be made very difficult, if not impossible, because food production will be unable to keep up with the needs of our vastly increased numbers, which will have caused massive deterioration of the environment. Or, our species will exercise control over its activities and accommodate them to nature. Except for a few lonely voices, this last option is presently ignored. It is not even discussed. Our society is like a rudderless ship charging ahead, going wherever it happens to go, while its passengers, mesmerized by its ever-increasing speed, cheer it on. We seem uninterested in assuming responsibility for ourselves.

We might expect democratic governments, above all, to be interested in human destiny, but this is not the case. They are preoccupied with another problem, maintaining themselves and keeping those who now run the country in control.

Ideas and inventions, once out of the bag, are on their own. We do not take it upon ourselves to see that they are used to benefit humanity. Instead, we become their servants—sometimes their victims. Powerful forces within us find ways to use them for instant gratification, to make money, or to harm each other. For example, automobiles were invented as a way of making people more mobile. They kill and maim large numbers of people, guzzle irreplaceable natural resources, pollute our air, and have done irreparable damage to our cities, causing them to spread and consume vast amounts of productive farmland. Because automobiles have so changed our environment, people travel greater distances—and must now contend with congestion and gridlock—so that they do not get to work, stores, or school any faster than they did one hundred years ago. Television is another example. Once seen as a panacea for education, it has in fact become the opposite, fostering passivity, violence, sexual promiscuity, self-pampering, and "want-it-all-nowism" among young people, turning them into what Norman Mailer calls "vidiots."[4]

Because of the complexities the modern world they must deal with, people are often distant from problems society could do something about, so these problems are often overlooked. In primitive societies where people lived in small groups of extended families and clans the situation was very different. In the more humane communities, for example, there was a place for everyone. Both the young and the old were assigned tasks they could handle and made to feel they had a place. Today, it is considered normal that a portion of the population cannot find work. While we usually provide these people with food and shelter, we ignore their emotional needs. It does not occur to us that despite all the progress we have made, we ignore important human needs many primitive societies handled well. Now most problems are dealt with by specialists working within narrowly defined areas often established by political haggling. Solutions are directed toward specific goals, like building highways, and fall short of the overall good they could achieve. This leaves many problems unnoticed, unaddressed, or improperly resolved. No person or group is held responsible for uncovering or dealing with them.

Biologist Garrett Hardin's often-referred-to "Tragedy of the Commons" describes a classic mechanism that undermines accountability. The grass-covered commons was owned by a community to be used by its members for grazing their herds. It was obvious to everyone that the more animals one could have graze on it, the richer one would become. Because no one was responsible for damage done, and because the greatest abusers were the biggest winners, the commons was overgrazed and thus destroyed. Today, the "commons" comprises many things that people share that are inadequately controlled. It includes the seas we fish, the air we pollute, the natural resources we extract, and the aquifers we pump.

Unless we change and actively demand accountability, as over-population, food shortages, and hard times spread, the glue that holds society together will disintegrate and a philosophy of every man for himself will take over.

THE PROBLEM OF CHANGE

Change is desirable when it improves our lives or when it helps us avoid dangers. It may be necessary to introduce pollution control measures, for example, in order to bring about a healthy, more stable environment. In chapter 3 I discussed how individuals react to change. Let us now look at groups and human society.

Society needs conventions that support order and keep the behavior of individuals and groups reasonably predictable. Customs, values, and traditional ways of seeing things help do this by resisting change. Organizations such as churches, schools, governments, as well as the public may share beliefs that resist change. Sometimes there are good reasons for these beliefs; often, there are not. The United States' economy depends on the manufacture of increasing amounts of consumer products. We believe that it is desirable to do so. Changing this belief and reducing production would have a traumatic effect on our economy and would hurt many people—even though it would further our chance for ecological survival.

Forward-looking individuals who see a need for change are often opposed by the majority of people who are bonded together largely by intellectual lethargy, groupthink, and fear of the unknown. Those who step out of line to remind us of what needs to be done are usually ignored, marginalized, or persecuted. Instead of facing problems and solving them, society may ignore them or blame them on scapegoats.

Society's resistance to change is strong and its roots are deep. Ervin Laszlo notes that a society will resist change as long as the current environment supports the minimal level of stability needed to maintain its principal institutions, worldviews, and the values on which they are based.[5]

The resistance is enhanced because there is always a time lag between a stimulus and a response, and the gap can be relatively long when it involves an organization adapting to a changed reality. Because change often takes place rapidly in today's fast-paced world, such delayed adaptations may be inappropriate or even damaging. The likelihood of responding inappropriately predisposes organizations to resist needed change. Consequently, many contemporary organizations are ill-suited to society's needs. For example, the Catholic church is unwilling to deal with the world's exploding population or its own shortage of priests in practical ways. Schools are neither able to cope with the social problems they face nor willing to provide the education young people need to function in today's environment. Society often clings to beliefs, ideas, institutions, and ways of doing things even though better ways have been found and new information has been discovered, holding on to archaic methods even after conditions have changed.

Clinging to things as they are foregoes opportunities for improvement such as the adoption of the metric system in the United States. Indiscriminately refusing to change or adopt new ideas can be dangerous, such as refusing to move out of a flood plain after having experienced a bad flood. We as individuals combat change and society compounds this by adding its layers of resistance onto our own.

Our personal image of the world—how we want it to be—is often outdated; and society's image is even more so. The world that we see ourselves interacting with is gone. It only exists in our minds. Some people act as though they live in a world without nuclear weapons, and most governments act as though population and environmental problems are minor issues. Like individuals, societies resist rapid transformation, but embrace and passively accept many creeping changes without considering where they will lead. We are so enamored by new innovations in technology and business that we are blinded to the liabilities they may bring with them.

The modern idea that individual welfare and rights are preeminent means that they take precedence over the welfare of our species. This idea goes against social stability, and, as I explain in chapter 8, the constructive workings of evolution. As is obvious in the United States today, families are weakened when individual members see their own needs as more important than those of the family as a whole.

PERSISTENT INFLUENCES

A shared experience can impart a spirit that brings out the best or the worst in us. People visiting a zoo or a nature preserve are generally good-natured and treat each other well, but some sports fans or members of a lynch mob can be very nasty—even deadly. The Japanese occupation of Shanghai, anarchy in Somalia, civil wars in the former Yugoslavia, and the south Los Angeles riots moved some people to behave like monsters and others like saints. Movies and television attach a mystique and "glamour" to violence that is hard for some people to resist. By revering the rich and powerful, society motivates some people to try to gain riches and power by any means possible. We make little effort to influence our impressionable nature in ways that actually benefit society.

Directly rewarding people with money and prestige has an even greater potential to influence what they do. A young person who has a way with words may become an English teacher or an advertising copywriter. We have a shortage of good English teachers, but are well-supplied with clever copywriters. What would we like this bright young person to become? We offer low pay and little prestige to those who choose to teach, while those who persuade us to buy something we don't need are rewarded with high pay and membership in a country club. To carry out unappreciated tasks such as teaching and running government agencies, we rely on idealists willing to make personal sacrifice, and when such people are not available, we are content with incompetence. The way we remunerate people has little to do with their contribution to society. Some of the most highly paid people have a negative effect, such as those who produce violence-filled television programs for children, or lobby to lower environmental standards to benefit corporate interests.

Talent and ability in my own profession, architecture, is poorly utilized. Take two college classmates: One wants to make the world a more attractive place and has talents that would help her do so, the other wants to earn a lot of money and has a gift for doing that. The former becomes an architect, the latter goes into real estate development. The architect makes plans that are either ignored or drastically changed by the developer so that profits can be maximized. The configuration of our environment is a by-product of the developer's pursuit of profit. He ends up both making money and shaping our surroundings. The architect ends up with an ulcer. Talent that could be used to make our environment more beautiful and livable is largely wasted.

The same is true in many other areas. A huge amount of human potential to achieve good is lost because certain talents, such as selling or politicking, predominate over others more suited to the task at hand, and also because of mechanisms that come into play when people interact with each other. We will look at these and other specific societal problems in later chapters.

WE ORGANIZE

We want to think that organizations act more sensibly than individuals do. Organizations, too, have a sum total of knowledge, talent, and skill that is greater than that possessed by any member of the organization. But here too we are disappointed.

When people work together, structures and procedures must be established. Detailed records such as wages earned and amounts to be billed must be kept by businesses and government in order to clarify relationships between people and to keep people from cheating. Rules and penalties clarify and reinforce acceptable behavior. Such mechanisms are cumbersome, add work, decrease efficiency, and can negatively influence what we accomplish. The larger the group or organization and the more it does, the more complex its structure, which channels activities and information and formalizes procedures, must be. This enables it to do specific tasks—often very efficiently—but it also limits its flexibility and restricts its ability to do other things.

Within the structure itself, people interact, and new forces such as personal ambition, jealousy, and insecurity appear, interfering with efficiency and objectivity. Sometimes decision-making becomes so awkward and politicized, and so many conflicts arise, that the group's actions are convoluted reactions to its own political strife rather than objective problem-solving.

An organization works well when its goals and those of its members are the same. However, people view their organization differently. For some it may be a way to further a cause or profession, for others, a means to gain power, prestige, friendship, or a sense of importance. The use of an organization by its members to pursue personal goals can decrease its overall effectiveness and even pervert its purpose.

The way a group is oriented and structured rewards certain activities, discourages others, and affects how well actions are performed. It attracts, or simply includes, certain types of people as members and repulses others. It makes it easy for some types of people to gain leadership positions and hinders others. It determines the mind and the heart of the group or organization.

Like the human mind, all organizations are composed of parts—individuals, each with his own agenda, joined together to serve a common purpose. Over time, these parts evolve to work reasonably well together and form a unit that adapts itself to the environment. If not, the organization dies—more rapidly than is the case with biological entities such as ostriches or people.

Our Loyalty to Groups

We may join groups to achieve ends we cannot achieve as individuals, but there are other reasons as well—satisfying our need for security and desire to belong, for example. As we saw in chapter 3, our desire for approval can pressure us to conform; uncritically accept goals, beliefs, and activities; and cause us do things we would never dare—or want—to do alone. The group can even bestow legitimacy on grossly misguided actions.

Some groups and organizations take far more extreme positions than their members would take individually. Zealotry is contagious, and group membership obliterates many personal fears and doubts.

By belonging to an antiabortion group, some people become willing to go to jail for blocking entrances to abortion clinics. People who supported Hitler's "final solution" would never have done so or even suggested the idea if they were expected to act on their own. In March 1997 thirty-nine members of the Heaven's Gate cult followed their leader, "Do," in committing mass suicide. They believed that a spaceship located behind the comet Hale-Bopp had been sent to take them to a "level above human." History is filled with organizations and nations that have gone off in directions that can only be viewed as madness. At such times, levelheaded individuals who call for the use of good sense are utterly ignored.

Despite their display of solidarity, large groups (such as nations) have many cracks beneath the surface. Individuals and special interest groups within them each promote their own agendas. Cooperation between groups is even more tenuous, because areas of shared interest are deemphasized in favor of areas of competition. Also, in these intergroup relationships, ethical considerations are relegated to a minor role. To achieve meaningful cooperation between groups, areas of common interest must be found. There are people who see these problems and would like to improve the workings of organizations. However others benefit from the status quo and work to sustain it.

DANGERS WE CREATED FOR OURSELVES

Human culture has changed drastically in the past ten thousand years, and especially in the last one hundred. New types of organizations are operating in environments that are presenting them with ever greater challenges. We do not know how well contemporary civilization and its organizations can survive the next millennium—or even the next fifty years. Not only do we not know, we act as though we do not even care.

Human society, as an ecosystem in its own right and as part of the world ecosystem, has checks and balancing feedback loops that maintain a certain internal stability. However, our balanced relationship with nature has been destroyed. As will be discussed in chapter 10, damage may be inflicted on natural systems many years before the repercussions on humanity are felt. Much harm can be done before society awakens to what is happening.

We humans possess characteristics, such as greed, that may benefit us as individuals but that can harm human society or nature. Such characteristics did not threaten serious damage when our ability to inflict injury was more limited and our ecological niche was small. This has changed over the centuries. In the hands of a vicious man a chemical weapon can do far more harm than a club. A small amount of waste dumped into a local river by a band of Native American hunters could easily be absorbed—or might even nourish aquatic life. However, even moderate amounts of toxic wastes from modern chemical plants can wipe out entire species and significantly alter the ecology of a river. Individuals or organizations striving for their own well-being may, with the best of intentions, severely damage whole societies and even our entire planet as a safe, nurturing environment for human beings.

When nature is injured, all of humanity suffers. However, offenders believe that their own gain is greater than the cost. Consequently, many damaging practices continue to do great harm to nature before evolution eventually eliminates them. Cattle ranchers who profit by overgrazing on government land may irreversibly damage delicate ecological systems and soils before they are stopped by declining productivity or the government.

Are We Harming Our Genes?

With the best of intentions the human species also harms itself. Our humanitarian efforts may have turned bio-evolution to work against us. With the worldwide diffusion of means for controlling disease, clean water supplies, and our policy of helping the infirm, we are saving the lives of many genetically unhealthy people who formerly would not have lived to reproduce. Some of these people have harmful mutant genes that are now being incorporated into the human gene pool.

In many countries people with little education procreate at a high rate whereas the most educated may not even reproduce at replacement levels. In the United States this occurs by choice. In China, urban families, which include many of the most educated people, have one child, whereas some farmers and people in remote areas may have more children. The following purely hypothetical example illustrates the possible consequences of uneven population growth. Say segment A, 10 percent of the total, only reproduces itself by 80 percent every 25 years; segment B, 80 percent of the total, reproduces at replacement level; and segment C, 10 percent of the total, doubles every 25 years. After seventy-five years the population would have increased overall by 65.12 percent; 3.1 percent of the babies born would be descended from segment A, 48.45 percent from segment B, and 48.45 percent from segment C. (The similarity in percentages is coincidental.)

This would be good if segment C had the most desirable characteristics and segment A the least. However, as I noted above, in some countries the educated population is the least likely to reproduce in higher numbers. If the most educated are also on the average the most intelligent, and intelligence is an inherited characteristic, then the most intelligent people are reproducing at the lowest rate. This fact should, but does not seem to, disturb us.

If the least intelligent people are reproducing in vastly higher numbers than the most intelligent, there will be problems in the future. There may be both many untrainable people unable to find work and shortages of people capable of performing the ever-more-complicated operations needed to keep modern society running. Not being able to find someone capable of fixing a computer or a television set would be a nuisance; not having the brain power we need to educate our children, run the treasury, or operate nuclear power plants and storage facilities safely would be frightening.

Human society is rapidly approaching an unknown dangerous future. It is doing so unconsciously, by default. It has always been this way, but today there is a difference. In the past the laws of nature worked to protect us; now, as a result of our tampering, they may destroy us. In the following chapters we will discuss the roles that various sectors of our human community play in this. We will then look at what we can do about it.

5

The Psychology of Society

"Madness is the exception in individuals, but the rule in groups."
—Friedrich Nietzsche[1]

Gustave Le Bon once noted,

> The most striking peculiarity presented by a psychological crowd is the following: Whoever be the individuals that compose it, however like or unlike be their mode of life, their occupations, their character, or their intelligence, the fact that they have been transformed into a crowd puts them in possession of a sort of collective mind which makes them feel, think, and act in a manner quite different from that in which each individual of them would feel, think, and act were he in a state of isolation.[2]

Although psychologists and sociologists have studied groups, most of us know little about their behavior. This ignorance is becoming dangerous. In the following pages we will look at some things of which we should be aware.

CONFORMITY

Primates, including humans, have been around a long time. Conformity played an important role in this. As pointed out in chapter 3, our instinctive urges, ambitions, and desire for security press us to meet this need to harmonize. While fitting in is essential for social animals like us, it harms modern society in a number of ways.

When we are free to choose those with whom to associate, we like to join groups of like minded-people. We go even further and try to be like those around us. As members of a group we tend to align our beliefs to those of the group, avoid questioning or challenging basic tenets, and sometimes even condemning or attacking dissidents. Group assumptions become sacred, explanations for them are invented, and change is resisted. This behavior reinforces rigidity and stifles originality and new ways of thinking.

When damaging acts against people or nature are permitted or committed by our own government, groups we belong to, or our chosen peers, we are inclined to join these actions or condone them. Sometimes procedures, routines, and rules are even established for conducting such actions. They insulate individuals from the responsibility of making decisions and give such actions the appearance of legitimacy. Participants in activities like the use of torture or allowing companies to pollute can say, "I was just doing my job." Furthermore, when part of a group, people operate at a low level of intelligence; reason becomes irrelevant. Author Arthur Koestler wrote, "The group-mind must function on an intellectual level accessible to all its members: single-mindedness must be simple-minded. The overall result of this is the enhancement of the emotional dynamics of the group and simultaneous reduction of its intellectual facilities."[3]

In the past the groups to which people chose to belong consisted of families, neighbors, tribes, nations, and religious, political, and special interest organizations. Today there are new types of groups, ones in which participation is confined to passively accepting established ideas and practices. Such groups include fan clubs, television audiences, and generic collections of people lumped together by marketing and advertising firms. Though people's involvement with these groups is vicarious, the groups play important roles in many lives. To participate, one must conform by attending events, wearing insignias, assuming certain values, and using certain products. These groups promote consumption, produce a feeling of well being, and at the same time trivialize serious concerns.

Memes

The ideas and knowledge that we share and pass along, like everything else, must follow scientific laws—including those that govern evolution. Biological evolution is guided by the reproduction of genes. In his 1976 book, *The Selfish Gene*, English ethologist Richard Dawkins points to another method of perpetuation, the meme. Memes are elements of a culture or a system of behavior that may be considered to be passed from one individual to another by non-genetic means, such as imitation. Dawkins explains, "I think that a new kind of replicator has recently emerged. . . . Already it is achieving evolutionary change at a rate that leaves the old gene panting far behind." He goes on to provide examples ("tunes, ideas, catch-phrases, clothes fashions, ways of making pots or of building arches") and then explains memes further:

> Just as genes propagate themselves in the gene pool by leaping from body to body via sperms or eggs, so memes propagate themselves in the meme pool by leaping from brain to brain via a process which, in the broad sense, can be called imitation. . . .
>
> For more than three thousand million years, DNA has been the only replicator worth talking about in the world. But it does not necessarily hold these monopoly rights for all time. Whenever conditions arise in which a new kind of replicator *can* make copies of itself, the new replicators *will* tend to take over, and start a new kind of evolution of their own. [Emphasis in original.][4]

Ideas reproduce and evolve as they spread from person to person by word, example, or other means. Thus, it can be said that two systems of replication are now evolving side by side: DNA, which evolves relatively slowly, and memes, which by comparison evolve very quickly. These systems interact. Genetically inherited characteristics of our personality affect our choice of memes and our ability to pass them on. Memes can affect the types of mates we choose and thereby affect the gene pool. Herein lies a problem. It is impossible for DNA to adapt to the rapid changes now taking place in our culture. In other words nature cannot keep up with the transformations we bring about that very much affect it.

In his book *Thought Contagion*, Aaron Lynch furthers Dawkins's concept of memes and provides this example: "Memes emphasizing wealth as a criterion for recognizing a person's value create an intense drive to accumulate wealth."[5] People who have developed this drive do well financially and consequently gain power in society. In this way the meme reinforces itself and spreads. In a less direct way, memes that urge nuclear families to desire large, single-family houses on large lots have brought about the proliferation of such houses and the automobile traffic they generate along with other burdens they place on nature. Couples who have made the commitment to make these investments have a greater incentive to stay together and thus reproduce at a rate higher than less stable couples. The meme, and the environmental problems it causes today, proliferates.

TRUISMS AND NONQUESTIONS

From childhood on certain truisms—both actual and perceived—are imprinted on us and affect our thoughts throughout our lives. They are simplifications of reality that can save us much time when confronting new situations, things, or individuals. However, some truisms, such as "throwing money at schools won't help," are often inaccurate and can cause harm. Nevertheless, we like them. They allow us to carry on without thinking or feeling uncomfortable, and enable us to blame others when it is convenient to do so. We don't have to worry about the homeless, because "they like to live that way. If they didn't, they would get jobs."

Truisms have been institutionalized in nations, churches, and many other organizations. Asserting mindless patriotic truisms and hypocrisies is rewarded with acceptance and approval. We like to hear that our soldiers are the best in the world, and that no one can manufacture better products than our American workers. Real patriotism, on the other hand, is dangerous; it involves upholding basic tenets such as the freedom of speech, or opposing a person one believes is immoral. It raises important questions and may ask us to criticize our country—making us very unpopular.

Part of the job of politicians, journalists, ministers, and teachers is to proclaim insincere banalities, particularly on special occasions. They must eulogize the ordinary and even praise scoundrels because of their position or popularity. Truisms and empty words cloud our thinking and misdirect our energies from reality to whatever is safe and comfortable for us or may benefit their authors.

We are told over and over, and without question believe, that "more," "bigger," "faster," and "newer," are good, growth is essential, consumption brings happiness, and that science and capitalism, if left alone, will solve most problems. They will provide us with an inexhaustible supply of natural resources, food, and an ever higher standard of living. This will result in a stable, prosperous world. It is far easier to accept these assertions than to question them and feel obliged to change our lifestyles.

Like the truism, the "nonquestion" (a question that does not occur to us or that we consciously do not raise) is another device that helps us get through the day. Were we to examine all that is known about the world, we would find a lot that is unsettling. To avoid this dilemma, we bury our heads in the sand. What we do not know or recognize does not bother us, and our ignorance enables us to happily direct our attention to pleasant diversions.

We have learned this lesson well. We are able to live seemingly oblivious to questions that in the "real world" are all too obvious. In George Orwell's novel *1984*, which is about total control in a future police state, many people were turned into "nonpeople," erased from people's minds and written records. We likewise allow uncomfortable questions to become "nonquestions."

When should population growth be stopped, and who should do it? Where will the increasing disparity between the world's haves and have-nots lead? What should be done to avoid such a gap in wealth? What will we do when petroleum runs out? Is it ethical to use up the world's petroleum reserves in an instant of human history and leave future generations with a changed climate and little topsoil? How will our descendants on an overpopulated, depleted planet maintain very costly systems for storing nuclear wastes? Why are American lives worth more than the lives of people of other nations? Do fortunate, gifted individuals have an ethical obligation take some responsibility for the less generously endowed? Are all people, or any people, capable of governing themselves, and if not, what can be done about it?

We avoid such queries, opting instead to maintain our blissful state of ignorance.

THE BATTLE OF BOREDOM

Most of us are unable to be alone with our thoughts for more than a few minutes. We reach for a button to push or something to read in order to fill the silence. We get bored with ourselves, our acquaintances, our spouses, our work, and the world around us. Many famous authors, including Pascal, Voltaire, Goethe, Kafka, Hemingway, and Sartre, among others, have written about this pervasive problem.

While boredom afflicts us all, it varies greatly between different societies. Sociologist Orrin Klapp makes an interesting but not surprising observation: "It is an irony of progress that boredom should be high in countries that have the best of it materially, in terms like ease, comfort, convenience, leisure, satiation, low personal output, high expectations, mobility, electronic communications media—all the conditions so familiar to us."[6] In the United States, as material well-being increased markedly between 1960 and the late 1980s, the teen delinquency rate doubled, suicide and homicide rates tripled, and births to unwed mothers nearly quadrupled.[7] How ironic that when we have achieved our goals of comfort, security, wealth, and status, we are likely to become bored and get into trouble.

Our minds are totally dependent on input from outside. Experiments as well as the experiences of people who have been held in isolation show that a lack of stimulation is hard to endure. To a lesser extent this is true for many people in their everyday life. Those with little imagination and few inner resources need stimulation the most.

We can hardly say that modern society suffers from a lack of stimuli. Klapp makes an interesting observation regarding this: "Boredom is but one thread in the tangled skein of whatever may be wrong with modern life; but I think it tells of loss of meaning from overloads of information that are dysfunctional either because of too much redundancy or too much noise."[8] Today we have become desensitized to much around us and need ever more dramatic events and sensations to hold our attention. Psychoanalyst Otto Fenichel tells us that we most often try to relieve boredom by finding something to act upon us rather than something we can act upon. By doing so, we choose passivity and avoid the potentially more stimulating challenges.[9]

A look at ourselves shows that we are largely reactive. The majority of us like a job where we are told what to do, where we can work on problems set before us, or where routines once established can simply be followed. Many of us like our recreation to be intellectually passive too. We watch videotapes, go on guided tours and cruises, and have hobbies that demand little creativity.

Boredom is bad, but its consequences can be far worse. Self-centered people who ignore the problems of the world about them can easily become bored. They could find much more satisfaction by working on these problems or engaging in other constructive activities, but their selfishness or lack of imagination keeps them from doing so. Street kids turn to drugs, reckless sex, stealing, and murderous gang wars for stimulation. Unimaginative national leaders fill the void with conflicts that can lead to violence. Amoral rulers turn to war and other atrocities to gratify themselves and appease their vanity. Those of us who do not indulge in violence find seemingly innocuous ways to fill our empty lives, but in the end some of these ways are as damaging as war. We play golf, watch soap operas and sports, and have children. We buy cars, boats, and vacation condominiums. Increasing numbers of people jet to distant places for skiing or suntans. A round trip for two people from Chicago to beaches in Thailand consumes about 1,000 gallons of jet fuel in modern aircraft operated at 60 percent capacity. In the endless quest to amuse ourselves, we consume, waste, pollute, and overpopulate. In spite of all our efforts and all the costs and devastation that results, we remain quite bored.

OUR ADDICTIVE LIFESTYLE

We have become hooked on comforts, pleasures, and conspicuous consumption. Like someone addicted to drugs, we depend on obtaining increasing amounts of money, diversions, comforts, status, power, etc. We cannot withdraw from our addictions without trauma. When the rate of growth slows, profits decline, people lose jobs, bankruptcies increase, tax revenues decline, money for the arts and sciences becomes scarce, people protest, and politicians do not get reelected. To keep growth increasing and to remain internationally competitive, governments are tempted to lower environmental standards and eliminate many of the amenities that make society humane.

VARIOUS VIEWS OF CHANGE

Human society must change to improve its condition, meet new situations, and avoid serious threats such as environmental destruction. While constructive changes are needed, stability and tradition are needed as well, otherwise societies and individuals would find it hard to interact with each other and would have little to depend on.

Individuals vary in their attitudes toward change. Some desire change for its own sake, some resist it almost on principle, and others evaluate each case on its own merits. We separate ourselves into self-reinforcing groups whose labels indicate our attitude toward change: "radical," "resistant," "self-protective," and "outer-protective." Note that the term "conservative" does not appear here. It has been omitted for a very specific reason: "Conservative" implies protecting and conserving, but we use the term loosely and imprecisely. Many of the people we call conservative are interested in conserving little besides their own property. We do not use this term to describe people who want to preserve our planet as a viable environment for life. Such individuals are more likely to be called alarmists, unrealistic do-gooders, or troublemakers.

"Radicals" are ideologues who are unhappy with the way things are and may want to change almost everything, perhaps for the sake of change itself. Sometimes they have little to lose by doing so. They are likely to throw out the good with the bad and replace one flawed system with another. By doing so they create much misery and destroy important self-regulating systems in the process. The 1917 Russian Bolshevik revolution lead to the deaths of millions of people, and destroyed an economy that would have eventually grown to meet the needs of the Russian people.

Those who are "resistant" have an instinctive desire to preserve society much as it is, largely ignoring its problems. They see change as a threat to order and security. Stability can become an obsession with them. This group tends to see events as largely isolated from each other. When the economy slows down, their overriding concern is to restore its expansion. They fail to see that the environmental exploitation and unending population growth now required to maintain a stable economy will significantly alter the future. They are so determined to resist changes to the status quo that they do not consider the consequences of their resistance. To maintain at least the appearance of the stability they so desire, they often oppose changes that are badly needed, such as reducing fuel consumption.

One change they fear greatly is a change in trends. They do not understand that because of the continuation of trends, each day is different from the preceding one. Today there is less topsoil, rain forest, and protective atmospheric ozone, but more pollution to contend with than there was yesterday. They do not understand that the growth they feel comfortable with will in time change most everything they value. Their attitude can be equated to being disturbed because a car you are riding in ceases to accelerate, without realizing that it is headed for a concrete wall.

The term "self-protective" describes people whose regard for their own well-being overrides everything else. They may see themselves as well-situated financially, politically, or in their occupation, and view any significant change as a threat to themselves. Some would like to revive elements of the past such as low taxes, inexpensive domestic help, and minimal government intervention in their businesses and affairs. Nevertheless, they very much do want change—creeping change that increases their wealth or their power, or raises their standard of living. They see a slowdown in the construction of shopping malls or the sale of new cars as a cause for alarm. For self-protectionists, as with resistants, slow-moving change seems normal —the way things ought to be.

Self-protectionists generally pursue political and economic power which gives them not only influence but the respect and admiration of their fellows. They associate with people like themselves and may send their children to prestigious conservative schools. This reinforces the attitudes of their class and insulates them from questions, new ideas, and innovation.

"Outer-protective" describes people who sincerely try to understand problems and find the best solution for them. They want to conserve and maintain order wherever it exists, and make changes needed to restore order where it does not. They want to keep what is beneficial and improve or change what is not. They can achieve much good. However, when concentrating on details without seeing the whole or when using bad judgment, they can create serious problems for both natural systems and human society. By eradicating certain "undesirable species," wolves for example, in order to help farmers, outer-protectives enabled deer populations to grow out of control. Hungry deer now invade farmers' fields and still many starve to death.

Different groups dominate society or a nation at different times. They affect how nations and people interact with each other and with the planet. At times one or two groups may become very committed and gain strength and dominate. During the Great Depression, organized labor had a strong influence on government. More recently, the influence is in the hands of Christian fundamentalists and businesses. When John F. Kennedy said, in his 1961 inaugural address, "Ask not what your country can do for you; ask what you can do for your country," many people backed altruistic activities and programs such as the Peace Corps and protecting the environment. We are now on a different track.

IRRESPONSIBILITY AND LAZINESS

Many of our problems, and the lack of good solutions for them, arise from irresponsibility, dishonesty, and laziness. Not many people would dispute this. Let us look at some examples.

My grandparents had a strong work ethic and belonged to a culture that made them proud of their stamina and their willingness to make sacrifices in order to adhere to their beliefs and achieve their goals. Today we want to forego sacrifice and our commercial environment encourages us to do so. Contemporary American culture, which is spreading across the world, places little value on self-discipline or self-reliance. Comfort and self-indulgence are unabashedly "in," and we vote for politicians who support these values. We like leaders who focus on easy problems with quick payoffs and ignore the more serious and thorny ones. When possible, we leave problem solving for tomorrow.

Confronting our most serious problems, such as the safe storage of nuclear wastes, can always be put off. The needs of today may outweigh everything else. Many farmers in Madagascar, Nepal, Mexico, and Brazil know that they are causing terrible damage by destroying rain forests and cultivating mountainsides, leading to massive erosion, but they find no other way to sustain themselves. Instead of learning how to preserve the topsoil and the forests, they do what is easiest. Significant areas of our planet are being permanently damaged this way, and the problem is rapidly growing worse.

Ignoring unpleasant problems and truths or discrediting the people who draw our attention to them provides relief and enables us to do what we want with a clear conscience. In recent years some businesses have deliberately confused issues such as public safety and environmental damage by trying to discredit professionally or personally those who have revealed these truths. For example, General Motors unsuccessfully hired detectives to find incriminating personal data on Ralph Nader because he made the world aware of safety deficiencies in one of their cars.[10]

For years some people have ridiculed Thomas Malthus's prediction that if unchecked, the world's population would in time outstrip world food production. They point out that farms have always been able to increase production faster than people's needs. That was partly true until around 1984, when per capita food production started falling behind. Simple logic should have shown these people that this would happen. However, Malthus's critics and those warning of other environmental threats prefer the logic of the fellow who is halfway down after a jump off the Empire State Building: "This is not as bad as I was told," he calmly remarks.

Blaming Others

The technique of blaming others in order to escape responsibility is used as effectively by organizations as it is by individuals. Governments at the 1992 Rio de Janeiro Earth Summit got off easy by pointing fingers at each other with respect to global warming. Developed nations demanded that countries in the tropics reduce the destruction of their rain forests and limit their rising emissions of carbon dioxide. These developing countries in return demanded that the industrialized nations reduce their levels of carbon dioxide production and relieve them of their overwhelming debts so that they would not need to cut their forests in order to make interest payments. In the end, all that was really achieved was much hot air and many stiff fingers. More specific commitments were made at the 1997 Kyoto conference, but we will have to see what happens before we can say that some specific results have been achieved.

On March 13, 1964, at 3:00 A.M., at least thirty-eight people watched from their windows in a New York City apartment complex as their neighbor, twenty-eight-year-old Kitty Genovese, was slowly stabbed to death. Over the half-hour period during which this took place, not one neighbor assisted her nor called the police. This seemingly bizarre behavior, and other similar cases, prompted two social psychologists, B. Latané and J. M. Darley, to investigate conditions under which people will or will not help others in trouble.[11] One thing they discovered was that when there are many observers of an activity, each individual is less likely to help than if there is only one or a few observers. Because so many people are present, we assume that someone else is helping, so our help is not needed. Or, we may feel that because no one is helping, it is all right not to help. Helping disrupts our own activities and can put us at risk, so if we can find an excuse not to help, we will use it. The level of responsibility of a group or crowd is generally less than that of an individual, especially one who is being observed.

When everyone else is wasting energy, we feel little compulsion to use it sparingly. After all, our efforts to save make little difference on a world scale and may handicap us in business or our profession. The vote of one individual makes little difference, but when most citizens do not vote the democratic process is quickly undermined. One person conserving makes little difference, but when most people do not, there goes the planet! How individuals might turn this tide is discussed in the final section of this book.

THE QUICK, EASY FIX

Some animals are genetically programmed to provide for their long-term needs. Squirrels do not save nuts for the winter out of wisdom, but because those that do so survive to produce descendants that save nuts. Because we lack genetic mechanisms that enable us to deal with the changing challenges we face, we must plan. As individuals, some of us plan reasonably well for ourselves and those we care about, but as a society our planning efforts are woefully lacking.

Society as a whole has little interest in history; demands quick, expedient fixes; and chooses leaders who promise easy solutions to complex problems. In times of war, or when war threatens, security takes precedence over everything else. In times of peace when business is slow or people are worried about inflation or recession, concern for the economy predominates. There are always many problems to be dealt with. The most serious threats to humanity and the planet must compete with these problems, which seem much more real to most people than some potential environmental disaster years in the future.

We cannot solve long-term problems unless we humans can act with consistency over an extended period of time. To do so we must feel relatively secure and be free of hunger, poverty, anxiety, mental illness, false values, erroneous beliefs, jealousy, and greed. As problems increase—and they will—it becomes harder for people to be concerned about long-term goals and make sacrifices for them.

We find ourselves trapped between a present in which people do not see the approaching dangers and a future where the problems that are closing in on us will be more difficult to resolve. There may come a time when we have decided to act, but the momentum of the approaching crisis will be too powerful to stop. As the problems we face become greater and more obvious, public pressure for change becomes greater. However, as people have trouble meeting their basic needs or desires and they become more despondent, their concern for the future and the resources they can use to protect it diminish. We are caught in a squeeze. As the likelihood of action increases in one way, it diminishes in another.

FICKLE SINGLE-MINDEDNESS

Human society, just like the individuals of which it is composed, is fickle. Not only you and I but the media, our leaders, and people in general give their attention to only a few things at a time, such as drug abuse, national defense, education, or taxes. After a while new issues draw their concern away. Issues are either in fashion or they almost cease to exist. This fickleness inhibits consistent progress toward solving a broad range of complex problems. Perhaps the combination of our fractionated minds, our inability to think of more than one thing at a time, our need for conformity, our resistance to what is new, and our propensity to be easily bored restrict both our interests and our ability to attend to the broad range of problems we face.

Narrowly focused thinking is as much a problem of the educated as the uneducated. For example, in 1962 I wanted to start a "survival center" where discussions regarding all threats to human survival would be held and literature related to the topic would be distributed. Those who became involved (including professors from the University of Chicago) were only interested in the international peace aspect of human survival. Thus, a "peace center" was established. Today, it is the environment that is "in."

Ideas and opinions, like clothes and music, are subject to fads. This can have serious consequences. In 1994 the extreme right wing gained control of the U.S. Congress. Without an objective evaluation of facts, the simplistic idea that government does everything poorly and business does everything well swept the country. Because this idea became so popular, politicians were afraid to oppose it. Consequently, to varying degrees, most joined in the dismantling or downsizing of important government programs and support for activities such as education, basic research, the arts, helping the unfortunate, and protecting the environment. Because most politicians went along with this and there was no strong opposition, a large percentage of the public also gave their support to the movement. Years of painstaking legislation and program building were quickly destroyed in the name of fiscal responsibility, with little analysis or understanding of what would follow. Perhaps one can say that the programs condemned themselves. If those that had been cut had not been successful, the general public would have been more aware of the problems and would have sought ways to resolve them rather than cutting the funding.

Like people, nations seem to learn only by experience. Unfortunately this can be expensive or even tragic. Waiting for a nuclear disaster to take place before taking effective measures to prevent it has obvious disadvantages. Nevertheless, experience remains the most effective means of group learning and initiating change. After a catastrophe, groups as well as individuals alter their behavior in order to avoid a recurrence. However, as we know, the desire to transform weakens with time. After World War I, the victorious nations established the League of Nations, which was intended to prevent future wars. As details were worked out, the League was not given the power its originators had intended, and when conflicts arose, its more powerful members disregarded its decisions.

When I was a child, most states made gambling illegal, for good reasons. Many families of compulsive gamblers were suffering, children were improperly fed and clothed. Some gamblers turned to embezzlement and stealing to pay their debts—and the activity itself attracted criminals. Today, without looking at why this was done, states are scrambling to get on the bandwagon to repeal these laws. Society's poor memory complements its fickle behavior.

PUBLIC PSYCHOSIS

Not all of society's destructive tendencies can be attributed to nations' usual goals or to their members' behavior. Just as leaders can be pathological, society can also have a sizable proportion of warped individuals who affect a nation's behavior, making it more paranoid, violent, or hedonistic than it might normally be.

English psychiatrist Anthony Storr, commenting on public behavior, wrote, "We may imagine that so-called normal people could never believe in anything so ludicrous as the delusional systems of the insane. Yet historical evidence suggests the opposite. Whole societies have been persuaded without much difficulty to accept the most absurd calumnies about minority groups portrayed as enemies of the majority."[12]

A bored, pleasure-loving society can develop a craving for perversion and cruelty. In ancient Rome, where slaves did most of the work, many citizens had little to do but seek amusement. They became hedonistic and acquired a taste for cruelty, such as that demonstrated at the gladiatorial games.

Society, like individuals, reacts in certain predictable ways. It presses people to conform, ignores reality, resists change, becomes dangerous when bored, succumbs to fads, is often fickle and irresponsible, and sometimes vengeful. Experiences such as wars, depressions, and luxurious living affect its thinking and behavior. Sometimes the pathologies of influential individuals lead a group to do mindless, grievous things.

History shows that we pay dearly when we overlook these facts.

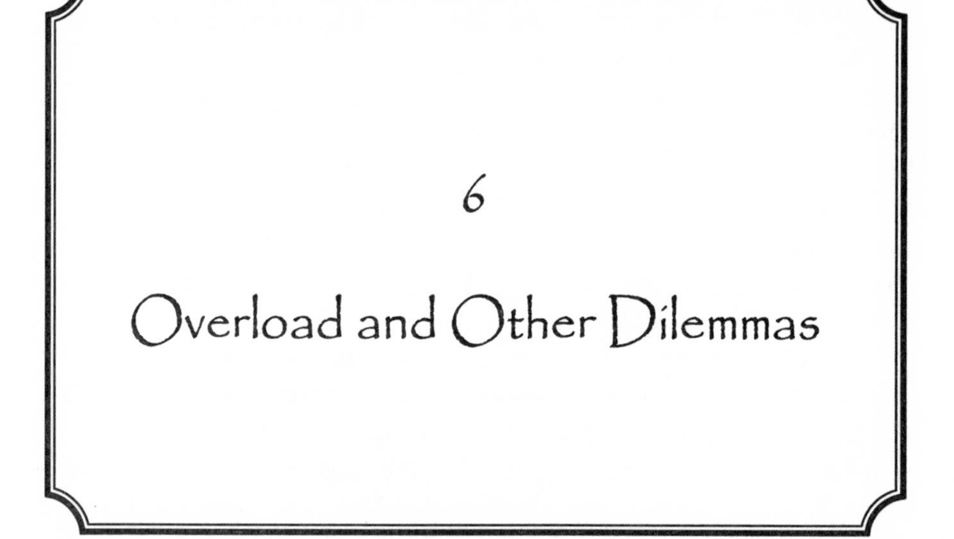

6
Overload and Other Dilemmas

More possibilities, more technology, more information, and more problems present the public with more things to know about, journalists with more to report on, and the government with more to manage. To cope, the government can set up more departments, enact more laws, and establish more programs. But we humans are stuck with our limited brains—and a desire to spend some time doing things that are pleasurable or just relaxing.

As I discussed in chapter 2, when individuals are presented with more problems or information than they can handle, they either become confused, nervous and ineffective, or indifferent and sloppy. They may resort to drink or TV. Today, human society finds itself in exactly this position, although it does not recognize it.

We are making efforts, sometimes even serious ones, to remedy some recently discovered problems such as endangered species. Nevertheless, problems are appearing at a rate faster than our ability to fix them. There are many reasons for this. Our troubles are now more numerous and complex than we are willing or able to deal with, and as discussed in chapter 2, evolution has not prepared our minds for this. This disparity was noted by Jose Ortega y Gasset in 1930: "The disproportion between the complex subtlety of the problems and the minds that should study them will become greater if a remedy be not found, and it constitutes the basic tragedy of our civilization."[1]

No one person can have all of the qualities and understanding needed to run a complex nation-state effectively. The problems of my own vocation can illustrate this. Once it was possible to be a good designer of buildings if one was artistic and knew how to build with stone and wood. Today an architect must be able to work with many new materials; have a grasp of structure, mechanical systems, acoustics, etc.; be able to manage complex projects; as well as be a businessperson and a salesperson. The architect must also be able to work knowledgeably and tactfully with others who insist that there must be more space here for an essential beam or there to run air ducts. I do not know anyone who possesses *all* these capabilities. Today it is virtually impossible to be a good all-around architect—or national leader, especially considering the limitations of the "leader personality" described in chapter 9. Nevertheless, this difficulty can be overcome, as we shall see in the final chapters.

Because of the way in which our democratic government now works, even the public, Congress, and the president must confine their attention to a few major problems, such as those of the budget or crime, at a time. Action on complex, controversial issues requires time-consuming public and legislative debate and carries political risks. Consequently, many important issues are largely ignored. Once we could get by in this way, but today, because of the number and magnitude of these threats, we cannot.

One would expect that the more difficulties we face and the closer we approach a catastrophe, the more likely society would be to face and solve its problems. However, as wars and economic crises have demonstrated, the very opposite may occur. A nation may procrastinate or turn to leaders who promise appealing but unrealistic solutions or a return to an illusionary past that is neither desirable nor possible to achieve.

Often the solution of a problem may be so unpleasant that we avoid facing it. Take, for example, the need to control atmospheric carbon dioxide. At the December 1997 summit on global warming held in Kyoto, Japan, a tentative agreement to restrain carbon dioxide emissions was reached. Already, though, conservative senators with an eye on votes say the agreement will never be ratified by the United

States unless changes—unacceptable to some of the other participating nations—are made. Even if the agreement were to move forward, however, it would only slow global warming, not stop it. It could also cause unforeseen economic difficulties around the world that would likely result in public demands to abandon the agreement.

Nations are unwilling to even consider curtailing the use of motor vehicles or the generation of electricity. Highly polluted China, for example, is determined to increase automobile production even though there isn't enough land now to meet current needs for streets and parking. Greatly reducing vehicular movement in our world today would leave many cities, especially in North America, helpless. Switching from fossil fuels to nuclear energy to generate electricity is exchanging one set of unresolvable problems for another. (Although harmful emissions such as carbon monoxide would be reduced, nuclear energy leaves us with the problem of waste disposal and storage—for thousands of years—as well as makes us vulnerable to horrendous acts of terrorism.) Simply raising these questions would be political suicide for an elected official.

When we are not consistently reminded of situations such as the loss of biodiversity, increasing discipline problems in school, and poorly stored nuclear wastes, they fade from our—and politicians'—minds. Concerned individuals and advocacy groups tend to direct their attention to no more than several issues, such as air pollution or inner city schools, leaving other wider issues and most importantly, broad overviews, unaddressed.

I was once actively involved in a futile attempt to expand the Cincinnati chapter of the World Federalist Association—an organization promoting law rather than war as a means for solving international disputes. In time we learned that its goal was too broad for most people. People prefer to focus on simple, specific goals such as outlawing abortion or preventing the construction of a dam, not ambitious, long-range goals that are even more difficult to achieve. However, limiting our concerns to those that are easy to deal with is not the answer.

INFORMATION OVERLOAD IN SOCIETY

Once human beings, like all other creatures, knew what they needed to know for their own and their species' well-being and survival; they depended on this information, could easily call it up, and used it. This is no longer the case.

While researching this book I soon learned that no matter how much time I spent collecting data, new information was being generated faster than I could locate or absorb it. If I tried to write a definitive book on this subject, it would be outdated before I was finished. Fortunately for me (and you), I am only trying to make a point.

Some people say that we are now doubling human knowledge every five to seven years. Language, writing, arithmetic, printing, libraries, and computers have each improved our access to this knowledge, and the "information superhighway" will advance it even further. However, these tools are not helping us as they should.

Vartan Gregorian, former president of the New York Public Library, said during an interview with Bill Moyers, "Unfortunately the explosion of information is not equivalent to the explosion of knowledge. So we are facing a major problem—how to structure information into knowledge. Because otherwise, what is going to happen? There are great possibilities of manipulating our society by inundating us with undigested information. One way of paralyzing people is by inundating them with trivia, giving them so much they cannot possibly digest it all in order to make choices."[2]

Urban planners become so immersed in data that they lose track of what it is that makes a community good. Governments and the soft and hard sciences suffer from a data glut, in which they often fail to find information that could help them. The people who produce this clutter are rewarded with recognition and promotions, so they continue to produce it.

Thus we are burdened with huge quantities of data to sift through. Sometimes we choose to focus on irrelevant material, sometimes we are forced to give it our attention. It becomes "noise" that makes it difficult to distinguish what is important for us from what is meaningless. Much of the data that fills our minds is trivia or relates to things we cannot affect, supplanting introspection, thought, and creativity and diverting us from dealing with many matters that are essential for our well-being.

Another problem is that we treat data differently according to how it suits our inclinations. We remember and store in active data banks information that promotes us in our profession or instructs us how to make money or weapons. We are also attracted to that which entertains or titillates. Other types of information may fail to reach the places where they are needed, and are consequently forgotten.

We, as individuals and as a society, simply cannot keep up with change and with the flow of information. We need to interact with the world we live in, but because we may base our thinking on incomplete or erroneous data, our actions are often directed at a world that no longer exists, the one we still see in our minds. Thus, our actions are often inappropriate.

We Don't Learn

The library where I did much of the research for this book covers a block and a half in downtown Cincinnati. Sometimes, after having discovered a book full of ideas that are new and wonderful to me, I stood looking at the rows of bookshelves and thought of how much knowledge and wisdom these books contain that could help us today.

There are books by humanity's greatest thinkers and by people who have painstakingly looked into and contemplated many facets of our physical, social, and spiritual life. All of these books have been read. People must have learned from them. But, as a society, we have used the wisdom contained in them poorly.

Perhaps the people who read these books were unmoved, have forgotten the lessons in them, misunderstood them, found no way to react, or were ignored by others when they did. Perhaps the trivial and erroneous simply drown out what is important.

With all that we experience and learn during our lives, we should become very wise, but we do not. Society, with all its accumulated knowledge, should do even better, but it does not. Georg W. F. Hegel put it in this humiliating way: "What experience and history teach is this—that people and governments have never learned anything from history or acted on principles deduced from it."[3] Sadly, we have not even learned this—yet. In the last chapters of this volume we will look at some ways we can do better.

The Implications for Democracy

The media are very limited in their ability to disseminate news. They are simply unable to report on the vast amount of happenings that the public should know about. Even if they could physically cover and present all the news, they are also limited by what we can and are willing to absorb. While newspapers and television were reporting on events in Somalia, equally terrible things were happening in other African countries such as Sudan. These events were largely ignored, giving the public the idea that the situation in Somalia was unique. During the time that much attention was given to global warming, the earth's topsoil was washing away, largely unnoticed by the media. When candidates are running for office, past actions in their careers that were highly meaningful to certain interest groups, such as people interested in education or the environment, but of no interest to the general public, may be totally ignored.

In a democracy, politicians need an informed public that demands responsible behavior from them. Today, no public citizen can possibly acquire all of the varied and detailed knowledge needed to adequately evaluate the performance of his representatives. The media cannot even give enough coverage to all matters of concern. If it could, the public would also need to have a thorough understanding of the background of all subjects discussed. For example, what are all the implications of free trade? Unfortunately, most of us are more interested in other things.

As I write, the president is about to give his State of the Union address. I suggest that we watch it on television. "Oh, no!" someone in the room, who considers herself to be a very good American, peals out with a sound of disgust, "there are game shows on that I don't want to miss."

The average family keeps their television sets on about seven hours a day, sometimes in more than one room at the same time. The set goes on when they wake up in the morning and when they arrive home from work or school. It stays on until they leave the house or go to sleep. Television offers diversion and crams our minds with sports information, the intimate details of celebrities' lives, and most important, brand names. Time thus spent is time not spent thinking about other things. In other words, we choose to distract ourselves with meaningless information, rather than focussing on and absorbing the information that will enhance our chances of survival. Increasing public ignorance of what matters as problems pile up makes it more difficult for us to meet the challenges that we face.

Special interest groups are not so easily diverted. They concentrate their attention on areas important to them. This gives them a decided advantage over citizens who feel the need to spread their concern over many issues—if they would take the trouble to do so. This factor built into democracies is shifting influence away from the people. Now, concerned citizens fight each battle as it arises, but fail to deal with the special interests themselves. We do not know where this will eventually lead. It is a problem we need to resolve.

MISSING CONNECTIONS

Problems, seen singly, are often quite easy to solve. But nature is not simple; things in the natural world are connected and interact with each other. In our simplistic attempt to improve our condition, we unknowingly influence systems that have interdependent parts.

When we set out to do something, along with understanding the task at hand, we should know what is available to help us, how our task relates to other things, and the possible side effects our contemplated actions will have. This is not always possible. Interrelating facts takes time, complicates tasks, delays solutions, and can be difficult. To simplify decision making, we often do not use even all of the limited information we do possess.

An interesting incident that demonstrates our failure to connect comes to mind. A physicist I knew built a solar house with many energy-saving features. He located it about twenty miles from where he worked and far from shopping and other essentials. I suspect that his family used considerably more energy for transportation than they saved by using the sun.

When people followed traditional practices, the side effects of their actions could be anticipated because they had become clear over time. However, since so much of what we do now is new, unforeseen side effects may appear. When freeways were first built in cities, there was no experience to guide us and little or no thought given to how they would affect metropolitan areas. The result was that we seriously damaged the physical and social structure of cities, destroying neighborhoods and causing urban sprawl. We dealt with automobile traffic as something unto itself, not as a component of greater urban systems.

To reduce auto pollution, some people call for more efficient, less polluting cars and better public transit. They do not look at what it is we are really trying to achieve by moving people and goods. Achieving the underlying purposes of this movement with minimal pollution might best be done by redesigning cities and by making better use of electronic communications so that people do not have to travel far to work, shop, or for recreation.

We find it easier to deal with symptoms than causes. By doing so, we get quick results, although at the cost of possible future problems and/or undesirable side effects. We save by constructing buildings cheaply, and then pay much to maintain them. To solve crime, we put more people in jail and keep them there rather than rehabilitating them; to improve education we build new buildings or put computers in classrooms rather than paying to provide better teachers and smaller classes; and to avoid war we build more terrible weapons rather than spending money to promote peace.

Most of the systems we deal with regularly are subsystems of larger wholes. Our banking system is part of our monetary system, which is part of our planetary ecological system (natural resources, forests, and food production have monetary value and interact with our economy). Manmade and natural systems interact with each other to form a larger system that encompasses them both. For example, some religious groups are helping to bring about the extinction of plant and animal species by hindering the dissemination of birth control techniques needed to restrain exploding human populations which are destroying wildlife habitats. In her book, *The Coming Plague*, science writer Laurie Garret describes how the loss of certain species will probably lead to devastating pandemics in the future.[4]

Economic models currently used by governments and businesses ignore nature and the distant future. They also ignore the negative effects of pollution, soil erosion, and resource depletion. Doing so simplifies the models and eliminates uncomfortable ethical problems such as concern for nature, future generations, and the poor in Third

World countries. Based on such models, the costs of extracting diminishing natural resources, of producing pollution, and of repairing the damages pollution causes to people and nature are counted as part of our Gross Domestic Product. It is factored in as the value of goods and services produced by society, even though it's for cleaning up the messes we create. So the more we pollute, the higher our GDP is! Because the individuals, organizations, and agencies that champion this thinking are powerful, highly respected, and well financed, people believe that these economic and policy models must be good.

Often the negative side effects of a technology (way of doing things) affect only people who neither benefit from nor have any influence on the technology's development. Consequently, their views are seen as immaterial. Poor people who live near polluting factories are victimized by the pollution even though they may not benefit from these factories (e.g., be employed by them) or have any control over them. Usable land on the edge of the Sahara Desert is turning into desert, possibly due to worldwide deforestation and the combustion of fossil fuels that cause global warming. Even if the people who live there understood this, there is very little they could do about it.

TECHNOLOGY ON THE LOOSE

The Pandora's box of new technology and ideas is being uncritically opened ever wider as we eagerly welcome all that floods out, believing that in some way it will be put to good use—although experience teaches us otherwise. Those who cry, "Wait! Think!" before we let it all out are viewed as alarmists.

Once the monsters are out of the box, it is usually impossible to put them back in. Throwaway packaging, mind manipulation techniques, and nuclear technology cannot be uninvented. Our wide open box may ultimately give us a world contaminated with chemical, nuclear, and cerebral junk and terrorists who can destroy whatever and whomever they wish.

Technological change is outstripping our efforts and ability to control it. In developing technology, our goals are clear, straightforward, and simple. In utilizing technology, our goals are open-ended (in this case open to unplanned possibilities) and value oriented, thus more difficult to clarify and agree upon. Without strong controls over the use of technology, it is difficult to foresee where it will lead us and impossible to direct its effect on us in positive directions.

While some individuals give serious thought to where technology is taking us, most people simply greet new developments with open arms. However, truly creative ideas and solid thinking, such as those of Galileo and Mendel, are often ignored. They may even be resented. The best educated and most intelligent members of society are often snubbed by government and the public. It is as if in a human being, the cerebral cortex were disconnected from the rest of the brain.

We make poor use of our potential. Ideas and inventions come from a very small number of intelligent or talented individuals who usually envision only positive possibilities for their creations. However, when these creations are able to make money or control people, there are opportunists waiting to exploit them. Some people saw the educational possibilities of television, others saw an opportunity to make money by vulgarizing it. The latter group succeeded.

Today there are many ideas and technologies that are being pushed to their limits with little thought of where they will take us. Unrestricted international trade, widespread jet travel, and workforce reductions do more than produce cheaper goods and increase profits. I will describe in chapter 12 how they can break down the fabric of families, communities, and cultures by producing rootlessness and insecurity and consuming huge amounts of natural resources.

THE PRESSURE TO SPECIALIZE

The human mind cannot understand, or work with the biosphere, technology, and society as the complex, interconnected whole they are. Our conscious mind, able only to consider one thing at a time, has a difficult time dealing with this.

To cope with the complexities that confront us, we dissect reality and place it into many small separate boxes that are easier to work with. Society rewards individuals who select one of these boxes and orient their training and thinking to its contents.

Thus we have come to rely on experts who are often very good at solving specific problems. This success, however, largely depends on isolating problems from the larger reality. Some highly skilled surgeons show little concern for the body as a whole when treating their patients, and like many hard-working, highly trained individuals, they may lack adequate knowledge of important issues that informed citizens should be versed in. Teachers have been taught much about how to teach, but too little about the subjects they teach and the things an educated person should know.

We are encouraged to learn a lot about a little, but not to gain a broad understanding of ourselves, our civilization, or our planet. Society employs, respects, and prizes experts, particularly when they can help others make money, win wars, or win elections. Generalists who raise disquieting questions and point out problems from a more global perspective are largely ignored and sometimes punished. It is far easier to become a successful expert than a competent generalist. All of this encourages us to become a society of individuals who are narrowly trained, each pulling on his own oar with no idea where he is headed. Psychologist Robert Ornstein and biologist Paul Ehrlich describe this problem:

> It is now easy, even desirable in terms of career advancement, for a young geneticist to know nothing about human development and behavior; for a physicist to know nothing about ecology; for a physician to know nothing about the lives and society of his patients. No Nobel prizes are given for understanding how the earth is being transformed, for knowing what the sum total of modern science means for society. No long-view, long-term understanding is prized when promotions are considered, and few appointments in universities exist for those whose knowledge does not fit a "slot" in an academic department.[5]

One example is economics, a field that should be all-encompassing. It is influential because governments and businesses base decisions on data supplied by economists. Economist Herman Daly and theologian John B. Cobb Jr. reveal some disturbing facts about the training of economists:

> A recent study of graduate education in economics concludes that "graduate economic education is succeeding in narrowing students'

> interests." According to the study's survey of the perceived relevance of other fields to economics, physics scored the lowest, and ecology or any other biological science was not even listed among the fields to be ranked. . . . Small wonder that economic models sometimes conflict with biophysical realities.
>
> Those students of the discipline who do raise radical questions about it are rarely appreciated, indeed, they find jobs are scarce and encounter difficulties in getting work published, They are likely to be denied a place on the program of guild meetings and to be made to feel unwelcome there.[6]

This concept is expanded upon by eminent political scientist John Herz, who criticizes academicians in his own field as being disinterested in the problem of human survival. In an address he noted that they spend their time "dealing with trivia—counting and rearranging the chairs on the deck of the *Titanic*. . . . Masters, in our time, still are chiefly those who pursue parochial objectives, interests that are parochial compared with those in global survival."[7]

We have produced actors who may perform well by themselves, but who will not work together. Human nature and our specialized groups encourage us to erect protective walls that others must not penetrate. Specialists develop their own jargon-filled languages, making communication with others difficult. Nature cannot be understood or effectively dealt with in this way. The specialization mentality creates coordination problems within disciplines as well as between them.

The success of the "specialist," whom Ortega y Gasset refers to as a "learned ignoramus," can lead to a dangerous arrogance.

> In politics, in art, in social usages, in the other sciences, [the specialist] will adopt the attitudes of primitive, ignorant man; but he will adopt them forcefully and with self-sufficiency, and will not admit of—this is the paradox—specialists in those matters. By specializing him, civilization has made him hermetic and self-satisfied within his limitations; but this very inner feeling of dominance and worth will induce him to wish to predominate outside his specialty.[8]

It must be noted that organizations, too, including governments, are specialists. They are structured to perform explicit functions, not to serve the overall needs of humanity. Their agendas thus become the aggregate of specific pressures which often conflict with the common good.

Specialization is necessary and has produced many successes for humankind, but at a price, one which we may not now notice, that gets higher and higher as time moves on. Human beings have circumvented many natural laws that maintained planetary order. However, we have not put anything in their place. Our destiny is now largely guided by the accumulated uncoordinated effort of technical, business, and political specialists, each working on details of his own specialty and not caring about the whole. We are thus pushed into the future by many single-minded decisions that fail to consider where we are going.

OUR TALENT FOR THINKING SMALL

We and our society evolved to think small, but the problems we have created for ourselves are very large. Journalists, historians, politicians, businesspeople, and the rest of us are myopic, seeing details but failing to see how these fit into a bigger picture. Like the blind men, we focus on the parts but not the whole elephant. We see the logic of economic growth, the balance of power, deterrence, and civil defense, but we fail to see the absurdity of the whole that we have put together.

In projecting the future, leading analysts rarely consider how trends in areas such as electronic communications, population growth, medicine, climate change, terrorism, and as yet unknown factors interact. By neglecting to consider interactions and by failing to draw people's attention to them, our forecasters fail to alert the public and governments to situations that could be ameliorated or perhaps even prevented entirely.

We are unable to rationally balance the difficulties (economic, social, political, etc.) of reducing carbon dioxide emissions against the risks of global warming. There are things we could do to cut emissions, such as stopping urban sprawl, that we don't even think about, largely because we do not ask the right questions. To make our country more competitive, some people suggest saving money on education, research, and libraries without considering the future effects of doing so. It is grotesque that older workers, some highly educated and competent, cannot find work when businesses are "streamlining" (reducing their workforces) in order to increase profits. We talk about these different but related problems, but seem unable to see them as parts of a whole in need of sensible, comprehensive solutions.

Not only do we have trouble thinking on a large scale, we often fail to focus on what we really want to accomplish. Consider the problem of heating houses efficiently: The useful question is not how to better insulate homes, but rather how to enable people to be comfortable and function well with a minimum expenditure of energy and money and minimum damage to the environment. Stating the problem in this way opens up a broader range of possible answers. Part of the solution may include new methods of heating and designing clothing that keeps people comfortable in cool places. Poorly formed, narrow questions eliminate many good ideas! Conversely, a well-formed question is far closer to being answered, since we know where to look to begin formulating solutions.

Early humans must have seen their greatest dangers as arising from nature. Droughts; long, cold winters; and hurricanes could be devastating. As people settled down and multiplied, they increasingly perceived hostilities from other humans to be the greatest threat to their security. While the danger of an all-out war should not be taken lightly, now a new category of danger looms over the others: the breakdown of the planet's life support system. Ironically, the natural dangers we now confront are not the ones faced by early humans, but ones we created ourselves. Most people do not understand this. For them the major enemies remain people: other nations, religions, criminals, welfare recipients, capitalism, communism, government, or environmentalists. It is as though they spend their time worrying about termites as their house goes up in flames.

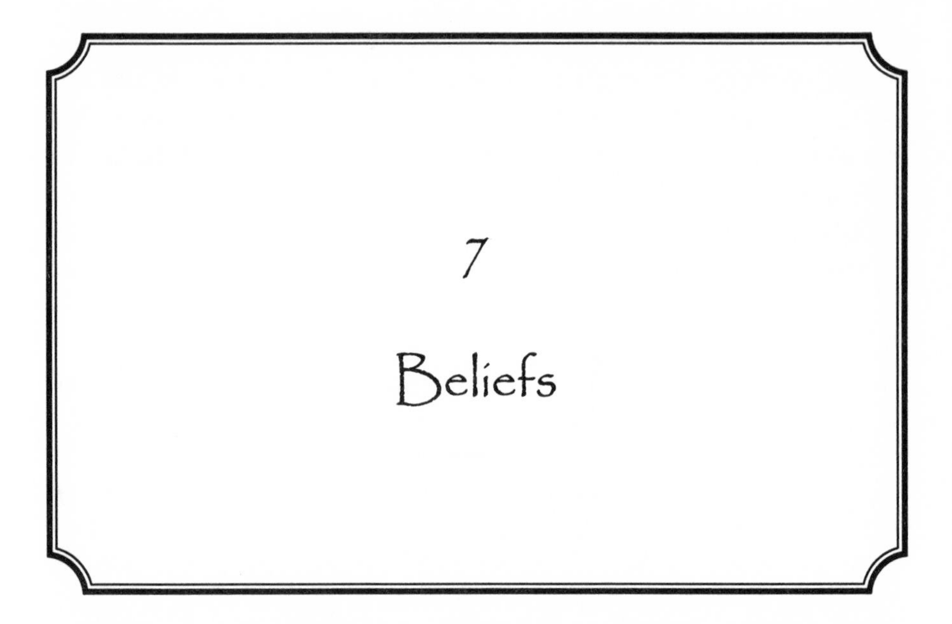

7

Beliefs

Straightforward models of the world that are based on what we perceive leave much unsaid. Why did a drought kill our crops? Why did our baby die? What happens to us after we die? Without answers, the world seems senseless and our lives meaningless. Nearly all of us seek reasons for what is happening to and around us. For some, not knowing is terrifying. Without explanations for such events, we lack a consistent foundation for our behavior. To satisfy this rudimentary need, we *believe*—we accept many things as being true with or without evidence that justifies our doing so.

Beliefs are necessary for peace of mind and to interact constructively with the world and each other. We must believe that the police or fire department will respond to our emergency call, that our bank will still be in business tomorrow. Our models of the world would be very incomplete if they relied only on what we could observe. The earth would probably be flat for us if we did not believe what others told us. History would be a blank—and so would science, unless we had conducted experiments ourselves. Often it does not matter whether a belief is true or not as long as it helps us to function.

In primitive cultures beliefs survived from generation to generation if they promoted stability. Otherwise, in time, they and sometimes the culture that held them became extinct. Because evolution favored individuals with the ability to believe, this ability became an important part of us. Like everything else about us, our belief system evolved to further our survival, not to provide us with truths.

OUR EAGERNESS TO BELIEVE

Science has given us insights that help us better understand the universe, our planet, and ourselves, but many things remain a mystery. Thus, in spite of considerable evidence, there are people not satisfied with, or even perturbed by, scientific findings. The popularity of astrology, gurus, and faith healers attests to this.

We want simple, satisfying explanations for what affects us. This may lead us to see random events as being part of a hidden order. We see connections where they do not exist. It takes a strong, and far too rare, respect for truth and reason to overcome this craving—along with a willingness to accept unpleasant facts and one's own ignorance. We favor arguments that support appealing beliefs and discredit opposing evidence. If it appears to further our interests, give comfort to or impress others, we eagerly seek out, recall, and interpret evidence and ideas that support our personal beliefs.

Jeremy Campbell describes our eagerness to believe:

> We have a built-in propensity to accommodate to almost anything the world may throw at us, no matter how bizarre and disjointed it may be. The mind leans over backward to transform a mad world into a sensible one, and the process is so natural and easy we hardly notice that it is taking place.
>
> We can turn nonsense into sense because that is the way the brain has been designed for a world where a fast, plausible interpretation is often better than a slow, certain one.[1]

Human experience, and tests conducted by Michael Gazzaniga,[2] show that when we cannot, or will not, see a rational link between data, our need for an explanation compels us to concoct a fictitious one. Thus we construct a system of beliefs based on relationships, real, or if necessary, invented.

Gazzaniga cites a case where the connections between the left and right brain of a patient had been surgically severed. The left hand is governed by the right brain, so in this instance, all connections between the left brain and the left hand were destroyed. Nevertheless, the patient, prevented from observing his left hand, used his left brain to explain what his left hand was doing. He was compelled to invent reasons for his action even though the left hemisphere could have no idea what the right hemisphere was directing the hand to do.

Invention can take another interesting twist. James E. Alcock, a professor of psychology at York University in Toronto, points out, "We are not always able to distinguish material originating in the brain from material in the outside world, and thus we can falsely attribute to the external world the perceptions and experiences that are created within the brain."[3] Our need to believe is so strong and so tilted toward accepting explanations that promise desired goals, that we eagerly accept questionable data. Once beliefs are established, we may forget the data on which they were based. This leaves us with strongly held beliefs we can no longer substantiate or question. These beliefs can close our minds to arguments against them and to data that could reveal their errors. Though wrong, such beliefs become our reality. We may be willing to fight and kill for them.

Julius Caesar wrote, "Men willingly believe what they wish." What they wish might be what their peers or a persuasive leader believe, what they were bought up to believe by their parents or an institution, what fills their emotional needs, or simply what provides easy answers. A belief is seldom a conclusion reached after an honest intellectual search. It bothers few of us that the world's people hold widely differing religious beliefs, and thus most, if not all, of them must be wrong.

Espousing erroneous beliefs can bring success, whereas a better understanding of reality can at times put one at a disadvantage. Christopher Columbus benefited Europe through his stubborn mistaken belief that he could find a short route to India by sailing west from Europe. Occasionally battles are surprisingly won by generals who underestimated their adversary's strength. Too often excellent projects that have every hope of being successful are abandoned because possible difficulties are anticipated. Conversely, we can all think of a number of worthwhile projects that we would not have undertaken had we anticipated the unforeseen adversities we were to encounter.

BELIEF-MAKERS AND GROUPTHINK

With the rise of civilization, humans came to rely on chieftains and priests, both of whom had access to information unavailable to others and were able to set goals and solve problems. Most people believed what they were told by their leaders. Today we still have a tendency to accept unsubstantiated assertions if they are made by powerful politicians, the media, or by famous or charismatic individuals. Our beliefs are also strongly influenced by our peers. We think that if most people believe something, it must be true.

As I described in chapter 3, peer pressure is a very strong force. Most people would rather accept the opinions of their peers and of those they admire than think for themselves. Conforming makes one more popular and acceptable; questioning is uncomfortable and risky. Members of a group or social stratum assume a package of beliefs, prejudices, and values that they hold in common. A political conservative accepts one package, a socialist another.

People who belong to groups generally cultivate the idea that their group and its beliefs are better than those of others. This is true of nations, tribes, religions, social classes, and often even occupations. Doing so strengthens the group, protects beliefs, and gives members self-confidence and a feeling of worth. However, it also spawns conflicts and closes minds to ideas that could give the group a clearer picture of reality.

We are also prisoners of our time. Whether we realize it or not, we assume many of its beliefs, prejudices, and values. Dr. Ervin Laszlo writes, "Each epoch in the past believed that its values and beliefs were right and eternal—rooted in the nature of reality, in divine will or in human reality—yet each epoch managed to evolve values and beliefs in tune with the prevailing conditions of human existence. The current sophistry of regarding modernism as an ultimate expression of human nature is no more valid than corresponding assumptions in the past."[4]

Beliefs, truisms, and nonquestions are often communicated indirectly, without speaking. We assume and do not question beliefs such as growth is good; or that big, new, and fast are better than small, old, and slow. Some believe that those on welfare are lazy, while others are convinced that corporations conspire to bilk the public. Some people believe that a large colonial house in an "exclusive" suburban subdivision is a very good thing. In some groups, one does not question whether the Bible is the word of God or the purpose of reciting the Pledge of Allegiance daily in classrooms.

By questioning, we jeopardize our credibility and our position among our peers. Drawing attention to the fact that the emperor's new clothes are no clothes at all can draw the ire of others. Accepting mutually held truisms and avoiding nonquestions makes life simpler, more comfortable, supposedly safer, and relieves us of thought and responsibility.

This reticence to question precludes learning and excludes much meaningful data from our models of reality. In many groups environmental problems and the increasing disparity between the world's rich and poor fall into this category. Many of us show little concern for the depletion of important mineral reserves. It is treated casually or not at all in our schooling, therefore we do not believe it is important, and because we do not see it as being important, it is not taught in many schools. This downward spiral perpetuates our problems by narrowing our input and observations further and further.

What is not mentioned is perceived as having no importance. However, when assertions are repeated over and over, or associated with a popular person, we tend to believe them.[5] In this way, unfounded facts can be introduced into our minds and accepted as genuine.

WE WOULD RATHER BE "RIGHT" THAN CORRECT

While changes in beliefs have been creeping into human society as a result of the environmental and peace movements, the change has mainly been among people with little political or economic power. Sometimes the necessary realistic beliefs seem to be gaining ground and making progress, only to disappear when a new administration gains power. Until President Ronald Reagan was inaugurated, public awareness of the need to conserve energy, reduce pollution, and promote international order was increasing. Reagan, who had little interest in these issues, dampened public concern for them.

A clear, accurate understanding of the world around us should give one an advantage in solving problems and in competing with others in life. However, we are not as eager to accept facts as reason would suggest. People will often tenaciously hold on to useless or dangerous unproven medical treatments. Some people stubbornly cling to communism, even though the seventy-year experiment in the Soviet Union has shown it to be a disaster.

Refusing to believe irrefutable facts or failing to seriously consider likely probabilities can also be a problem. In spite of conclusive statistics, until very recently, many people did not believe that cigarette smoking could cause cancer. I think that both Presidents Reagan and George Bush were unable to believe overpopulation poses a serious threat to humanity, or that automobile exhaust and smokestack emissions really could cause forests to die and the world's climate to change. Their inability to believe allowed these problems to intensify and has significantly delayed finding solutions to them.

We cling to falsehoods, defend them, and often persecute those who question them. Usually, the more someone challenges our beliefs, the more determined we are to keep them. We are more interested in being obstinately "right" than in being in line with reality.

We may become so entrapped by dogmatic beliefs that we grow unwilling to examine evidence that counts against them. When I was an adolescent taking religious instruction, I was told that if I lost my faith, I would be eternally damned to horrible suffering. If I really had faith, I would not want to consider anything that challenged it; such consideration in and of itself would be a lack of faith. I was making a bargain with God. By accepting what my church told me and declining to use my mind, I would be guaranteed a place in Heaven. What an effective way to protect believers from reality!

Even without being instructed to do so, many believers impose intellectual isolation on themselves. They fear having their illusions punctured, preferring fantasy to truth. As Professor Alcock notes, "This belief engine selects information from the environment, shapes it, combines it with information from memory, and produces beliefs that are generally consistent with beliefs that are already held."[6]

Why are we so often disinterested in reality? One reason is that we want to retain the beliefs we acquired early in life. They are deep within us. To change a belief is an admission of having been wrong and probably having made mistakes. We have made a bad investment in persistently pursuing a bad idea and don't want to admit it.

COMPARTMENTALIZATION AND CONFLICTS

Michael Gazzaniga has observed that we often alter our beliefs to suit our needs or desires. He notes an experiment in which students were given an exam on which they had to do well and were given the opportunity to cheat. Those who said cheating was bad before the exam and yet cheated on it said that cheating was not so bad when asked the same question after the exam.[7] Seemingly moral people regularly justify questionable activities of organizations with which they are affiliated. An advertising executive will defend advertisements promoting consumerism as being essential for our economy and way of life. An executive in a fast food chain has ready arguments to defend the use of disposable cups and plates. Sometimes such people may knowingly lie, but I am sure many have convinced themselves that the activities of their organizations are both necessary and moral.

The compartmentalization of our minds and the difficulty some parts have in communicating with others enables us simultaneously to hold conflicting beliefs. It also enables us to act contrary to our beliefs with little discomfort. A Christian artillery officer may sincerely believe in obeying the commandment "Thou shalt not kill," and at the same time believe in the death penalty and also feel it to be his duty to bombard a village filled with civilians if so instructed by his government.

Compartmentalization also enables us to conveniently turn our problems into "their" problems. Reasoning that "guns don't kill people, people do," allows one to fight handgun control with a clear conscience. This is not stupidity, but willful blindness to connections that are obvious to unbiased people. Society is also compartmentalized. Normally there is little communication among people with different religious, ethical, or political beliefs. Consequently, there is little possibility for interchange, modification, compromise, and intellectual development.

Often we belong to several groups that have conflicting beliefs. With little effort we accommodate ourselves to each group while we are in contact with it. Many regular members of churches that teach Christ's doctrine of sharing with the poor support political groups that propose to cut taxes for the rich and reduce spending for social needs and education.

Conflicting beliefs can be an advantage for a group. A tribe believing it to be normally wrong to kill, but that it is all right when their chief orders them to attack their neighbors, has a decided advantage over tribes that either deem it permissible to kill anyone or feel it wrong to kill anyone. The former would kill each other; the latter could not defend themselves.

In the past humanity was able to survive conflicting beliefs. Groups were separated from each other and when they fought their weapons inflicted limited damage. If one group was wiped out, there were others left to perpetuate our species. Today it is possible that a major conflict brought on by irreconcilable beliefs could wipe out our species.

Like the blind men describing an elephant, we all see the world from our own limited viewpoint, and each of us instinctively thinks his own image of the world is accurate. Often it is the people who know the least who are most insistent at proclaiming their view as the correct one.

OUR BELIEFS AFFECT OUR BEHAVIOR

Beliefs have inspired people to invent tools, develop science, and create great works of art. Belief in God moved Bach to compose beautiful chorales and Buddhists to erect magnificent temples. Belief in the dignity of the individual led to the development of democracy. The belief that a cure for polio could be found led to the development of a vaccine.

As we know, beliefs affect behavior. For example, some people think that money is the measure of a man's worth and treat others accordingly. Women were executed in seventeenth-century Salem, Massachusetts, because they were believed to be witches possessed by the devil. Some of us believe people are basically good while others think of them as basically selfish and deceitful. The former approach new acquaintances with trust and a desire to cooperate, the latter with suspicion and hostility. Nationalism indoctrinates people with the simplistic idea that their nation is sacred, good, and superior to others. As a child I believed that our leaders and bureaucrats knew what was best and could solve all problems. Many people seem not to have outgrown this idea. Together these beliefs make it relatively easy for an egomaniacal leader to plunge his country into hostilities and atrocities.

Some people believe that an orderly, homogeneous world without conflicting ideas is possible and desirable. This belief can lead to war, "ethnic cleansing," or George Orwell's *1984.* Human creativity and the evolutionary process, which require variety and mutation, would be stifled in the regulated world these people envision.

We have inherited beliefs, ideas, and ways of thinking that are no longer suitable. Ideas of sexual, racial, ethnic, and nationalistic superiority are widespread. Many people believe that the only way to maintain peace is to have a greater number and more powerful weapons than one's adversary. We cling to many outmoded ideas that prevent us from pragmatically dealing with reality. But the content of our beliefs is not our only problem.

In spite of the weak foundation that supports their beliefs, many go to great ends to sustain their own ideas of what their faith asks of them. In some cases this goes as far as terrorism and even killing and torturing opponents. Many people have written about the dangers of compulsive believers. For example, Andrew Bard Schmookler states, "If Nietzsche is correct—and the conduct of our true believers would certainly seem to validate his assertion—then intolerance is the inevitable companion of deeply held belief, and holy war an inevitable cost of having a planet of impassioned believers."[8]

On the other hand, a lack of belief can be dangerous too. It can stymie necessary action. Vice President Al Gore gives an example: "Perhaps the most serious threat to our stewardship of self-determination—one that may be even more threatening than all of the others put together—is that so many people have come to feel that the process of change in which we are now swept up has gone so far and gained so much momentum that it has outstripped our capacity to affect it."[9]

Beliefs affect values and goals. A teenager who has just killed another to protect his turf asserts, "So I wasted him. No big deal." For him it's not. What is a big deal is a pair of fashionable sneakers or a racy car. If we believe that appearance and having plastic surgery is more important than saving for our children's education, we will behave differently than if these beliefs were reversed. If we believe that both overpopulation and street crime are problems, and that of the two street crime is the more serious, we will ask our politicians to concentrate funds on fighting crime.

The weight society gives to its different beliefs has a profound effect on how it conducts itself. The behavior of journalists, politicians, businesspeople—most people—indicates that they really believe that economics, money, "success," winning, power, and the like are the "real world" and matter more than the environment, our children's future, cooperation, love, creativity, and the search for truth. As a result, we are obsessed with our economy while the exploitation of people and nature continues. Historically, the importance people have placed on such beliefs has produced unfathomable human misery.

Modern industrialized society values what it sees as relating to itself. People with political and financial power often believe that only the world of politics, economics, and at times social problems is worthy of serious attention. Some socially concerned, well-meaning people believe that nothing has value except as it relates to people. Social programs, world peace, and education are important, protecting the environment is a secondary concern. There are social workers, politicians, businesspeople, academics, journalists, Christians, socialists, and conservatives who fall into this category. By focusing on human problems and ignoring everything else, we are blinded to the dangers engulfing us as a result of our relentless destruction of natural systems. We fail to see that the effects we have on the environment will one day have a direct effect on us.

Thomas Berry, Passionist priest and director of the Riverdale Center for Religious Research in New York City, feels that this problem is greater in our culture than at any previous time:

> During this period the human mind has been placed within the narrowest confines it has experienced since consciousness emerged from its Paleolithic phase. Even the most primitive tribes have a larger vision of the universe, of our place and functioning within it, a vision that extends to celestial regions of space and to the interior depths of the human in a manner far exceeding the parameters of our own world of technological confinement.[10]

OUR DETERIORATING RELATIONSHIP TO NATURE

In spite of what we may like to think, chimpanzees are genetically more closely related to us than they are to apes. Ninety-eight and six-tenths percent of our DNA is the same as theirs, and although our brains are different in size, they are similar in structure. Nevertheless, many people, fundamentalist Christians, for example, refuse to recognize this relationship. They classify chimpanzees as animals along with worms, even though they are genetically very near us and far from worms. They consider humans to be above nature and condone the human ownership of animals as they once condoned the human ownership of other human beings.

As we in industrial nations spend more and more of our lives in an artificial environment we lose contact with nature and come to think that technology and a thriving economy are the foremost foundations of our well-being. As a result, we show little concern for what we are doing to our planet.

For many centuries, Western civilization has been trying to free itself from the unpleasant constraints of nature. Most recently, public sanitation, antibiotics, sophisticated surgery, jet airplanes, computers, air-conditioned high-rise buildings, chemical fertilizers, and chicken factories appear to have helped us move toward this goal. In our minds, freeing ourselves from nature's limitations is good, and we have had successes in doing so.

We have, at least for the time being, consigned nature to a source of food, resources, and recreation—on our terms. Economic laws take precedence over natural laws, which we now largely ignore. In fact, we deny any rights to nature and have largely excluded it from legal protection.

People have not always held nature in such low esteem. Many Native American tribes viewed things differently. They saw themselves as being part of nature. Chief Seattle is credited with having said, "The earth does not belong to man, man belongs to the earth. This we know. All things are connected."[11] Smohalla of the Columbia Basin Tribes is quoted as saying: "You ask me to plow the ground! Shall I take a knife and tear my mother's breast? Then when I die she will not take me to her bosom to rest."[12] Beliefs such as this gave many Native Americans an environmental ethic.

Judeo-Christian beliefs, on the other hand, have given Europeans and North American whites an exploitive ethic. In the book of Genesis God ordered Adam and Eve and their descendants, "Be fruitful and multiply, and replenish the earth, and subdue it: and have dominion over the fish of the sea and the fowl of the air, and over every living thing that moveth upon the earth" (Gen. 1:28). In the seventeenth century, Latitudinarian* divines such as Archbishop John Tillotson and Bishop Edward Stillingfleet argued that God expected people "to make use of metals, stones, timbers, plants and animals, to advance the sciences, to increase the trade and prosperity of their countries, and in doing so to glorify their creator. When people built beautiful villages, country houses, cities, and castles, God was pleased with their industry."[13]

While the Christian ethic tells us that the earth is ours to use, it really places value on departing from it. The evil earth is a test; it is actually redemption out of earth into heaven that counts. Thomas Berry notes, "While none of our Christian beliefs individually is adequate as an explanation of the alienation we experience in our natural setting, they do in their totality provide a basis for understanding how so much planetary destruction has been possible in our Western tradition. We are radically oriented away from the natural world. It has no rights; it exists for human utility."[14] Many Christians believe that our planet is God's concern, not ours. He will provide for it, and when he chooses, he will destroy it. To worry about the planet is to distrust God. Although any rational view of population growth statistics is terrifying, the Catholic church forbids the practice of artificial birth control methods. Population growth is seen as God's concern and whatever happens, is His will.

Enter "Nature the Machine"

By the mid-seventeenth century machines such as wind- and watermills and clocks had become a common part of European life. European

*A liberal religious "denomination" that favored self-interest and acquisitiveness which arose in the seventeenth century after the English Civil War.

thinkers who had heretofore viewed the world as a living organism developed a new concept of reality based on the belief that nature is made up of modular interacting parts that operate like a machine.

Carolyn Merchant, a science historian at the University of California, Berkeley, thinks many of our beliefs that damage nature come from replacing organismic philosophies with mechanistic ideas that were developed in the period of Enlightenment. She writes,

> The rise of mechanism laid the foundation for a new synthesis of the cosmos, society, and the human being, constructed as ordered systems of mechanical parts subject to governance by law and to predictability through deductive reasoning. A new concept of the self as a rational master of the passions housed in a machinelike body began to replace the concept of the self as an integral part of a close-knit harmony of organic parts united to the cosmos and society. Mechanism rendered nature effectively dead, inert, and manipulated from without.[15]

The thinking of men such as Copernicus, Francis Bacon, Thomas Hobbes, René Descartes, Isaac Newton, and Gottfried Wilhelm von Leibniz made modern science possible. They reconceptualized reality as a machine and introduced ideas that have led to the subjugation and destruction of nature by humans. Their ideas have made the modern world possible, but they also removed controls that maintained ecological balance between humans and other species. They have permitted the irresponsible exploitation of our planet, which until recently, was considered a positive endeavor. Today, some people recognize the folly of this. However, for the most part they are people with little influence, and the momentum that must be overcome seems irrepressible.

Berry writes of people who put themselves above nature and prided themselves in their realism, "Little did these people know that their very realism was as pure a superstition as was ever professed by humans, their devotion to science a new mysticism, their technology a magical way to paradise."[16]

Late nineteenth- and early twentieth-century accomplishments in science, technology, and business led people to believe that the future would be bountiful. They had produced impressive results; they enabled Western culture to predominate over other cultures. But in our veneration of these achievements, we have largely refused to see their negative side effects, although they are now becoming ever more obvious.

The belief that science, technology, and the pursuit of self-interest will solve all problems and provide a steady stream of wonderful things nicely alleviates the need to act responsibly. Science and technology *could* do much to mitigate the threats we are facing—if we used them wisely. We have not so far. Why do the proponents of salvation by technology think that somehow, out of the blue, we will act differently in the future?

We Know What's Best for Earth

In the late sixteenth and early seventeenth centuries Francis Bacon introduced a new vision of the relationship between humankind and nature: People would benefit if humans controlled nature. Carolyn Merchant writes, "Much of Bacon's program in the *New Atlantis* was meant to sanction such manipulations, his whole objective being to recover man's right over nature, lost in the fall [from paradise]." Before, nature was accommodating and people did not need to work.[17] This idea, as well as the book of Genesis and past scientific and technological successes, gives us the cavalier idea that we know what is best for nature and have been given a manifest destiny to rule it.

Our recent technological prowess has put us in a position to powerfully affect our planet. In doing so, we are replacing a finely balanced mechanism that took millions of years to develop with a crude, fractionated, uncoordinated collection of actions that we do not fully understand. Yet, in our ignorance, we arrogantly see ourselves as improving nature.

Capitalistic nations revel in the superiority of the self-regulating market system over the artificial, centrally controlled communistic system. As with science and technology, the quick successes produced by market economies gives us a cockiness—a certainty that we are doing things right. In our rapture we fail to recognize that over the long term some mechanisms built into this economic system have harmful social and environmental consequences. I will discuss this in a later chapter.

We do not publicly recognize what really underlies our influence on nature. It is not our wisest and most responsible thinking, but rather the consequence of our drive to achieve comfort, fun, power, and riches.

BELIEFS CAN BE HAZARDOUS

Today many beliefs and practices are new, as our relationship to the environment changes from day to day. Under such conditions untested beliefs may be ill-suited to society's needs. Acting on these convictions can be hazardous because our impact on the planet is enormous.

Our Faith in Growth

People in developed nations believe that economic expansion is essential in order to maintain employment and the standard of living. Beyond this, we simply believe that growth—constant expansion—is intrinsically good. There is evidence (our concept of manifest destiny, the success of capitalism) that supports this view. Our experience shows that when growth stops prosperity stagnates and jobs disappear. Simplistic thinking and this experience pressures people in power to ignore alternatives to growth.

Nearly all small towns and medium-sized cities want to grow, but they fail to notice the effect that successful growth has had on the quality of life in large cities. They think they can continue to grow while escaping the drawbacks of big cities.

It hardly seems possible that we do not consider that at some time growth must come to an end for the sake of future generations, and for ourselves in the coming decades—and the sooner the better. In 1972, the Club of Rome, an informal group founded in 1968 to further understanding and to draw the attention of policymakers toward the interdependence of disciplines in the world system, sponsored a report entitled *The Limits to Growth*, which became an international best seller.* In order to downplay the book's success, detractors misinter-

*Donella H. Meadows et al. (New York: Universe Books, 1972).

preted theoretical projections as serious forecasts. They were then able to shrug off the very meaningful message of the book—that exponential growth has limits. In 1973 the English economist E. F. Schumacher's book *Small Is Beautiful** was published. Schumacher contends that economics should recognize both the basic and spiritual needs of people as well as the limits of our planet. He advocates small organizations and simple, appropriate technologies that meet these needs better than modern society does. The book has simply been ignored by most people, including those with political or financial power. If we really would seriously consider putting an end to growth, our answer would undoubtedly be "Tomorrow."

People in many developing countries know that their populations are already far too large and see their situations as catastrophic. Many people in developed nations, however, fail to understand the stark implications of overpopulation. They take issue with desperate efforts to control population, such as the one-child-family plan of China.

In spite of their problems, people in the developing world also want economic growth and a higher standard of living—and we have no right to tell them they may not live like us. But the world simply cannot support the consequences of the direction we are headed—having a population much larger than it now has with a standard of living higher than we now enjoy in the West. In fact, data shows the earth cannot even sustain the burden we are currently placing on it. Our unquestioned, undiscussed belief in the need for perpetual growth is driving humankind to ravage the surface of our planet like a swarm of locusts.

As I write, many businesspeople, politicians, and nations are promoting the idea that unfettered world trade is intrinsically good. Undoubtedly corporate executives, stockholders, and consumers will benefit. The total world production and consumption of food and goods will certainly increase. But many workers, communities, local cultures, and the planet as a whole will be losers.

The need to move more people and goods over greater distances will mean an increased use of fuel and higher levels of pollution. Relaxation of emission controls gives offending nations a competitive advantage, which rewards pollution. This should not be welcomed at a time

---

*(New York: Harper & Row, 1973).

when our planet is threatened by global warming largely caused by increasing amounts of carbon dioxide in the atmosphere. An expansion of monoculture (planting the same crop year after year) will make it impossible to return many nutrients and organic materials to soils. This will further deplete the soil and cause lost nutrients to become pollutants in waterways where they promote algae growth and the loss of oxygen. The relationship between people and their immediate environment will deteriorate even further than it has already. Simply put, our enthusiasm for free trade is driven by economic considerations, period!

While endless growth can only lead to some form of disaster, in itself it is hollow. Andrew Schmookler makes an interesting point: "Our cultural fetish of economic growth is a clue that our consumption is not nourishing us, and our national cult of growth is a sign of the triumph of hope over satisfaction, of the life of promise over the life of fulfillment."[18]

Paralleling our belief in growth is our perception that consumption brings happiness. Down deep most of us know that this is not true, but on the level of instincts that guide many of our activities, we seem to think this is so. Our pursuit of growth and consumption places a heavy burden on the environment and diverts resources and our attention from areas more in need of them.

The idea that a simpler life can be fulfilling is a nonquestion. Formerly, people made do with much less, but they did not seem to be less happy for it. Considering this possibility now, however (i.e., making life simpler), is simply out of the question.

WE PREFER YANG OVER YIN

Chinese philosophers saw reality as a process of cyclical change between opposites, which they called *yin* and *yang*. These were not not seen as separate, but as extreme poles of a single unity. Now, instead of respecting each pole of opposites, although we profess otherwise, most of us believe that masculine, aggressive, competitive, and rational (i.e., yang) qualities are superior to feminine, responsive, conservative, cooperative, and intuitive (yin) qualities. These beliefs are reflected in how we relate to each other and to our environment.

As yang qualities tend to prevail over yin when they compete, efforts to cooperate usually succumb to competition and aggression, propagating yang qualities. Our society worships winners and this affects our beliefs. For many people, winning means more than being right. We tend to believe statements made by a successful executive, the winner of an election or war, a popular entertainer, or a professor at a prestigious university. And our society places a higher value on competition than cooperation, on specialization than on generalization, and on business than on education or liberal arts. Competition ranks high in the modern Western mind. We believe it to be a basic rule of evolution, not realizing that among mammals, cooperation often plays as important a role.[19]

Capitalist nations believe it is natural and right that land has an inherent monetary value that is determined by the market. Land developers use the word "improvement" to describe anything done to or upon land that removes it from its natural state. When land is cleared of trees and topsoil, then leveled and paved for parking, as is the owner's "right" to do, it is said to have been improved. Economic and legal concepts lead us to destroy the natural goodness of land, water, and air.

INDIVIDUAL RIGHTS & OTHERS

Those of us fortunate enough to live in a democracy are guaranteed certain "rights," such as the right to believe and say what we choose. In our minds, these rights may be extended to include the right to a free education or to have free medical care. How extensive these rights should be is a matter of much discussion among citizens, legislators, and lawyers. Until now, governments have only recognized the rights of human beings. Although there are now laws protecting some animals, plants, and ecosystems, we have not yet fully recognized the legitimacy of animal rights.

Evolution works to strengthen living species. Inappropriate qualities and harmful mutations are weeded out of gene pools by the death of weak individuals before they procreate or safely raise their offspring. In democratic nations we have fought against the unpleasant aspects of this principle with considerable success. We even have organizations that promote the rights of indigent or retarded people to have children, even though society may be unwilling to support or care for those children properly. For us each living human is sacred, his or her "rights" and welfare are paramount. In chapter 4 I explained how this works against the good of our species and the "rights" of future generations.

Our current concept of "rights" raises what should be an obvious question: Is it more important for us to have as many children as we choose, treat them as irresponsibly as we wish, and exploit the planet, or would it be better for our children to inherit robust genes, caring parents, and an earth that is as healthy as the one in which we now live? Only we can decide, our children cannot. This is another question we do not ask, a nonquestion now answered by default. To ask it would be most uncomfortable for us. Insisting on our rights while ignoring those of future humans (as well as those of other species) is thoughtless, confused, and immoral.

What we may believe constitutes a right often creates other problems. Historically, parents, teachers, and religions provided children with uplifting stories and role models. They still try, but their influence (that of the parents as well as the role models) has been undermined by dazzling distractions such as television.

Governments license the use of broadcast bands to a limited number of companies and grant exclusive franchises to cable TV firms, giving them access to a block of our children's time that is larger than that of their parents, teachers, and church combined. These companies have a constitutional right, with few limitations, to broadcast what they wish—and most of us uphold that right. Their legal responsibility is to their stockholders—to make as much money as they can. To do this, they do what attracts viewers and then present them with as many commercial messages as allowed or as those viewers will bear. They use vulgarity, violence, and sex to attract large audiences.

I cannot help but believe that TV viewing does have a significant influence on people's behavior, including children's ability to concentrate on schoolwork. We find ourselves in a terrible bind. By committing ourselves to upholding the right of free speech, we open ourselves to a force that degrades the values of our society. I will discuss the ethical aspect of this in chapter 8.

BELIEFS AFFECT HUMAN RELATIONSHIPS

Newborn children are only aware of themselves. In time they learn that there are other people and other points of view, but for some years they think that everyone sees the world just as they do. As they develop, they learn that other people view things differently, but the children still think that only they are right. In adolescence, as Robert Ornstein writes, "a new ability of the self emerges, the ability to imagine and idealize things. Then the child characteristically believes that the idealized view is preferable for all: 'The world should be as I see it.' This view persists and becomes part of the different views we have of ourselves and others."[20] I think that all of us retain this way of thinking to varying degrees. In spite of the fact that our beliefs are often grounded on sketchy or questionable evidence and faulty reasoning, we are usually convinced that they are correct.

As a tool for survival, evolution has given us the idea that our family, tribe, gods, and beliefs are legitimate and good and that others are suspect and possibly evil. Individuals, societies, and cultures with strong beliefs are more likely to defend themselves than those whose beliefs are weak; they have a better chance to survive. This worked well in simple societies isolated from one another—and still does in many ways. However, today this protective mechanism can create conflicts leading to unnecessary hostilities and war, or it can deter beneficial cooperation.

Beliefs have often brought enormous suffering to humanity. Communism, the Crusades, the Inquisition, and Muslim zealotry are just a few examples. Religious beliefs have led to human sacrifice, persecution, banishment, wars, torture, and other atrocities. During the Spanish Inquisition, in the name of God sometimes more than one hundred people were burned at the stake in a single day. Frequently, one tribe or nation believes that land occupied by others is rightfully theirs. Nations with certain religious convictions often believe it is their duty to subjugate and convert infidels by force. Some people feel they are intrinsically superior to others, and that it is only right that they should rule inferiors. All of these mistaken beliefs have helped cause the turmoil that rocks the world today.

Most human beings have a strong desire for order. When groups with differing ideas of order come into contact with each other, serious conflict can result. Fascists expect others to conform to their ideas and values, conservative Muslims and some Christians want to create intolerant religious states, communists once wanted to use any means to convert the world to communism. Conflicting views between religious conservatives and scientifically oriented people as to the origin of our planet, ethics, and other questions paralyze our ability to solve many serious problems such as uncontrolled population growth. Antagonistic beliefs can make it impossible to agree on what the most serious problems are, and on how and even if we should solve them. Sometimes these distracting disputes reach such heights of pettiness as to make the old debate about how many angels can dance on the head of a pin seem almost sensible by comparison. Such confrontations use energy that is badly needed for solving real problems.

Dedicated extremists often have an impact on society that exceeds their numbers: They are rabid about their beliefs, so they become politically active, concentrating their financial aid and personal efforts on electing or defeating politicians based on one or two issues and ignoring all others. The pressure they exert has kept books out of libraries, made the teaching of creationism mandatory in some school systems, prevented sex education in areas where illegitimacy is high, and terminated U.S. funding for United Nations family planning programs. Fundamentalists of different religious faiths and political beliefs have rendered impossible any national or worldwide consensuses on many important questions and they consume the energy of those who would like to solve these problems.

In this atmosphere of confusion and confrontation, our endeavors to maintain peace, control the damage we are inflicting on the biosphere, and solve many other serious problems are narrowly directed toward individual problems, or at times, at only the symptoms of these problems. As a result, we are winning some battles but losing wars. Until we confront the underlying causes of the problems—which include ill-conceived images and harmful beliefs—we will have to rely on our luck for success. In later chapters, I will discuss how we can do better.

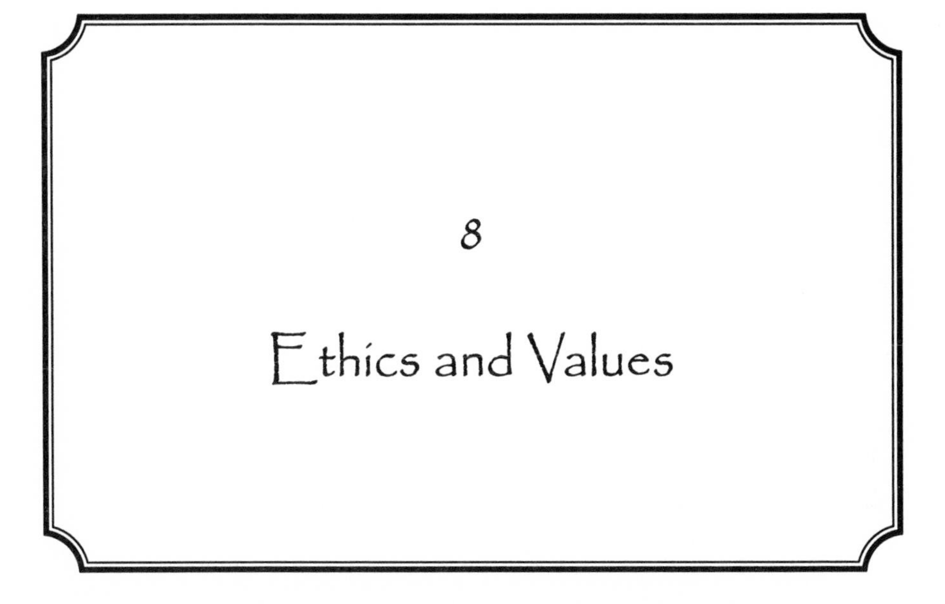

8

Ethics and Values

> "The most important human endeavor is the striving for morality in our actions. Our inner balance and even our existence depends on it."
>
> —Albert Einstein

Of the many obstacles confronting responsible behavior, the most difficult to overcome are probably those involving ethics. Even when we learn how to solve serious problems, we may simply choose not to do so. We may not care about people such as the powerless and future generations. And if we do want to act ethically, there are internal and external forces that influence us to do otherwise. Even when we succeed in following our moral principles, they may be such that they cause us to harm other people or nature itself.

ETHICAL SYSTEMS

Ethical systems are generally grounded on basic principles often seen as given by God or are based on what is considered humane behavior toward other people. While there are some widely accepted precepts, such as that one should not steal, there is no set of rules held in common by all societies and all human beings. In fact, some cultures have rules that directly conflict with those of other cultural groups. Sometimes interpretations of ethical rules may even lead to "holy wars" and genocide.

Some fundamentalist Muslims feel their religion compels them to kill infidels, nonbelievers who stand in the way of spreading Islam or who violate its values; they will be rewarded in heaven for doing so. Traditional Hindus believe that a person's situation is a deserved reward or punishment for behavior in a former life and therefore other people have no obligation to help the unfortunate, condoning and strengthening the caste system. Protestant and Catholic religious zealots fought each other over their differing beliefs in the Thirty Years War (1618–1648), one of the most brutal wars Europe has ever experienced.

Ethics That Apply to People Only

In chapter 7 I described how many Native Americans saw themselves as being part of nature. These beliefs encouraged them to respect nature and live in harmony with it. An unknown Indian chief said, "All things are bound together. All things connect. What happens to the Earth happens to the children of the Earth. Man has not woven the web of life. He is but one thread. Whatever he does to the web, he does to himself." Of the white man, the chief said, "He treats his mother the Earth and his brother the sky like merchandise. His hunger will eat the Earth bare and only leave a desert."[1]

Contemporary Western ethics is people-oriented. Many people of different faiths believe that our planet exists solely for the benefit of humans, and that there are no ethical laws describing how we should treat it. The Ten Commandments deal with people's relationship to God and to each other, but mention nothing about nature. A study made at Yale University found that in the United States, the more frequently people attend religious services, the less likely they are to show concern for nature.[2]

Many people whose primary aim is to acquire wealth see man-made laws protecting private property as sacrosanct. They believe it is their moral right to strip-mine or do whatever they want with land if they own it. Some "humanitarians" say that people are all that really count, that nature must come second. They do not recognize our roots in nature. They fail to understand that if nature is harmed, in time people will suffer.

Some people argue that birth control is wrong because it deprives potential human beings of the right to exist. These people do not consider that if the earth's human population is allowed to run rampant, the planet's capacity to support life will greatly diminish. Over time, if the population explosion is brought under control, many more people will have lived on a healthy Earth than on one that has been seriously damaged. In other words, in the long run, birth control will increase the number of people who will exist rather than decrease it. A controlled birth rate will allow the earth to support humanity for a longer period of time than rampant population growth could ever allow. In this latter case, those who live during the later stages of a "wind down" (degeneration, the result of our failure to control the birth rate), will live in misery like many people now do in Chad, Bangladesh, and Haiti.

"Rights" and Legalistic Viewpoints

Many in modern Western nations believe there are inalienable human "rights" that are given by God or inherent in democracy. However, "rights" are a human invention; they do not exist in nature. In fact, as I discussed in chapter 7, they often undermine constructive evolutionary laws. Nature favors the survival of suitable genes, and ruthlessly weeds out weak individuals. In contrast, in democratic nations, the rights of individuals come first. As noble an idea as this is, it can cause serious problems for future generations.

While respect for "rights," as we interpret them, would seem to be based on ethical principles, in reality it is based on selfishness. It concerns itself only with the present generation of humans and ignores the well-being of future ones, as well as other species and the biosphere itself. Sometimes too, this respect for "rights" is extended only to people who belong to a specific nation or religious group.

In our formalization of ethical systems we have a tendency to become preoccupied with rules and details. In doing so, we lose perspective and common sense. Some people believe that a place in paradise can be earned by performing religious rituals regularly and correctly, and by strictly following rules such as not eating pork, requiring one's wife to wear a veil, or by uncritically believing whatever —no matter how irrational—is written in some book. Dedication to ideas and rules may become so intense that broader religious requisites such as loving fellow humans, caring for the unfortunate, and respecting life are ignored.

Throughout human history religious beliefs and ethical ideals have been twisted and at times turned into nightmares. High-minded concepts can be misinterpreted in ways that satisfy some of our basest desires. The teachings of religious leaders have been contorted and misrepresented, leading to the torture and death of many millions of people. (Sometimes religious teachings go so far as to call for the killing of nonbelievers.) The crusaders, the Catholic inquisitors, and Islamic jihad warriors became so obsessed with the activities or mere existence of "nonbelievers" that they killed many of them, often in cruel sadistic ways.

We do not seem to have learned from these terrible experiences. With the elimination of hostilities between the East and the West at the close of the twentieth century, the lid that had suppressed smoldering prejudices was removed. People let their most brutal instincts loose in order to achieve allegedly lofty aspirations such as the founding of "Greater Serbia," or to produce religious states ordered by Allah.

REASONS FOR BEING UNETHICAL

It is not always easy to act in an ethical manner. There is much that impels us to misbehave. Sometimes it is because we lack the imagination to empathize with others or to be emotionally moved by situations that are not close to us, or we may let our desires, fears, and prejudices color our picture of reality. Sometimes, we do unethical things because we don't know what is actually right or wrong, or because we are misinformed about aspects of a particular situation. We may be misled by the ideas of a charismatic person or group, or get caught up in a fad or the hysteria of a crowd. We may get so involved in a limited world, such as that of money making, that we form our ethical principles from a very narrow, one-sided position.

Human Instincts

Basic human drives such as competitiveness, desire for status, fear, and attraction to violence, for example, often work against ethical behavior. When they do they are powerful and hard for the intellect to overcome. Fear drives people to lie, to turn on their friends, and to persecute or attack others. It draws resources away from real needs and directs them toward ends such as excessive military defenses.

Our selfishness, resistance to change, and drive to evade what we see as unpleasant can work together. Suppose we knew that our grandchildren would live in a world that has double its current population, serious shortages of food and resources, a changed climate, many coastal cities flooded, and many food-producing areas without adequate water. Suppose we also knew that we could largely prevent this if we made some sacrifices such as limiting families to one or two children, conserving petroleum, driving less, reducing consumption of electricity by foregoing most air conditioning, and saving forests by significantly reducing our use of wood products. If we knew without any doubt that we could avoid this terrible outcome by making sacrifices, how would we and our leaders act?

This is an interesting question, because in fact information now available to everyone indicates that if current trends continue, some of these predictions are likely to come true and others are inevitable. If we were honest and looked facts straight on, we would find ourselves in a very awkward position. Ethically we would be required to do without luxuries we have come to love, but that our ancestors got along very well without. Our response has been simple: Let this be a nonquestion.

Greed influences many people, but its greatest impact on society comes through its influence on people with financial or political power. When they behave badly their avarice can affect us all.

We know that wealth, power, and fame are often achieved by

means we consider immoral, yet we tend to respect those who have achieved these ends. This only makes these goals more attractive to those who seek them.

WHEN WE ARE NOT WATCHED

For a while I lived in a coach house on an alley behind a row of elegant townhouses. The alley ran between Studs Terkel's famous "Division Street" and several of Chicago's finest hotels. People on the street which the townhouses faced were well dressed, soft spoken, and polite. It was different in the alley. It was a place to throw trash, relieve oneself, or get a "fix." On the street one was observed; in the alley one could do what one wanted. The people who rip pages out of library and telephone books and scratch graffiti on toilet stall walls certainly do exist, though most of us have never seen them. And studies have shown, for example, that men in public toilets wash their hands markedly more often when they are watched than when they are not. Universities that name buildings after patrons and symphony orchestras that list donors in their concert programs know what they are doing when awarding public recognition like this. Most of us behave better when we are under scrutiny.

Most of our lives are spent in the alley, rather than on the street, metaphorically speaking. Much of what we do is not observed and depends on self-imposed ethical restraints that may not be very strong. Add to this our belief that "everyone else is doing it" and "just a little bit more won't hurt," and the authority of ethical principles can be weakened beyond repair. Discreetly dumping toxic substances into public sewers, deciding to sell the natural resources on public lands, and delaying the implementation of strict emission controls becomes very easy indeed.

Our Fingers Point the Other Way

We are good at criticizing others, but poor at applying our professed ethical principles to our own actions. Likewise, we do not see our own responsibilities as well as we see those of others. Many people are quick to criticize "those welfare recipients who spend their money at the beginning of the month on luxuries and throw their garbage out of their kitchen window." These critics do not see that the rest of us are doing the same on a different scale. We are squandering the earth's resources as quickly as we can and throwing our garbage (pollution and waste) into the atmosphere, waterways, and countryside. This is far more serious than what those feckless welfare recipients are doing—or not doing.

THE SHAPERS OF OUR VALUES AND ETHICS

The values and the ethical principles we follow, and how strictly we adhere to them, are influenced by those around us as well as ourselves. First our family instills and teaches moral values. Then, as we get older, teachers, clerics, groups we join, the media, friends, as well as people we admire affect our judgments. In time we may question the values with which we grew up, become curious about other tenets, or formulate ideas out of our life experience and learning. However, most of us unthinkingly accept the prevalent views and attitudes. Even if we do look beyond, our limited knowledge and the environment around us still influence what we adopt as our own.

Today, as in the past, people want to be viewed favorably by their peers. To achieve this we try to fit in; we adapt or bend our ethical beliefs and values to what is acceptable to our associates. Robert J. Lifton's "doubling" and "numbing" described in chapter 3 help us do this. These behaviors are mutually reinforcing. If none of my peers or role models is concerned about wasting resources or attending to the global environment, it must be all right to do so. If I were to object I would stand out and might even offend them.

Mind Manipulators

Today many parents spend little time with their children. Many children now have little moral instruction, and due to pressure from both religious and nonreligious people, schools refrain from teaching ethics as they once did. Traditional influences are losing ground to commercial ones that are skillfully designed to attract and hold attention.

Those commercial influences—advertisers and the media, to mention two—have powerful advantages over traditional ones: access to much of our time, an understanding of what motivates us, and sophisticated techniques for manipulating our minds. They bypass our better selves, overcome our weak powers of reason, and appeal directly to our strong, primitive instincts of greed, fear, and vanity.

The entertainment and sports industries assemble large audiences to whom advertisers can present their messages. Billboards, the backs of buses, and the pages of popular magazines provide places to catch people's eyes. Today it is virtually impossible to avoid advertisers' messages. Most of us do not seem to mind.

I find pornography disgusting, but I am not obliged to view it, and it is doubtful that it demoralizes our whole society. Advertising is forced upon us, and its message of joy through things and the means it uses to draw our attention promote shallowness, materialism, and waste. The propagators of pornography are looked down upon, as they should be, but advertising is a respected profession. Once it was possible for parents and community elders to monitor and select most of the information to which children were exposed. Today this is impossible. The mass media present young people with a constant barrage of superficial values, violence, and personalities who become their heroes and role models. Instead of presidents, explorers, scientists, artists, and saints, they are rock stars, actors, and athletes. Their values and lifestyles rarely encourage ennobling thoughts, kindness, cooperation, self-discipline, or responsibility. However, touting them is good for business, whereas promoting social virtue sells few products.

The Effect of Manipulation

Endless messages appealing to our most primitive desires cannot but have a negative impact on our mores. Advertising impresses on us the idea that happiness comes from owning an expensive new car. In order to afford it, we must work harder to earn more money. This can mean spending less time with our children and abandoning them to the care of the television set. The children's values and mores are then formed by default—as the side effects of efforts to sell by people whose goal is simply to make money. We do not seriously consider that there might be a connection between the time children and adults spend watching frivolous programs on television that tell them they must have, have, have, and short attention spans, passivity, low academic achievement, dysfunctional families, drug abuse, teenage pregnancies, and crime.

Years ago, Erich Fromm wrote of advertising, "As a matter of fact, these methods of dulling the capacity for critical thinking are more dangerous to our democracy than many of the open attacks against it."[3]

OVERLOAD AND THE DULLING OF OUR CONSCIENCES

As I discussed in chapter 6, we are now overloaded with possibilities, problems, and messages. Lacking full knowledge and time to contemplate, it is difficult to distinguish between the trivial and the important. Our sensitivity to many ethical issues has been numbed. We yawn at a television account of the bombing of a city, or an interview with the vice president is interrupted by an advertisement for a product we really don't need. Our newspaper headlines a story about an umpire strike across the top of the front page and relegates an account of mass starvation in an African nation to a small piece at the bottom. "Because of all this we cease to be genuinely related to what we hear. We cease to be excited, our emotions and our critical judgment become hampered, and eventually our attitude to what is going on in the world assumes a quality of flatness and indifference. In the name of 'freedom' life loses all structure; it is composed of many little pieces, each separate from the other and lacking any sense as a whole."[4]

CONFUSED GOALS AND VALUES

I have taught architecture at three different universities in the United States and one in China. Some of my American students had automobiles, their own apartments with many amenities, and traveled during their vacations. The Chinese students lived six or eight to a room in crowded dormitories with bare concrete block walls, no heat, and unlit hallways. They owned little. The most privileged American students were no happier than the Chinese, and more to the point, they did not learn more.

I had a friend who suffered for many years before she died of multiple sclerosis. I always enjoyed visiting her, because in spite of her condition, her zest for life, her courage, and her positive outlook always sent me away feeling good. In contrast, many affluent people in good physical health lead miserable lives and some even commit suicide.

All of this may neither surprise nor seem strange to you. Most of us know that if we are healthy, free, not deprived of necessities, and have challenging tasks to do, additional possessions do not increase our happiness. Recent research at various universities has shown this to be true.[5] Nevertheless, we allow primitive drives to prevail and passionately devote ourselves to the pursuit of goods, power, or fame, often sacrificing our families, overall happiness, and personal integrity in the process. This hurts not only ourselves, but human society and the planet as well.

ETHICS IN BUSINESS

Before the Protestant reformation in the sixteenth century, most trade was carried out by producers such as farmers, cobblers, and weavers. Their mores allowed them to make a reasonable profit for the work they performed. They charged a fair price for their product or service rather than the maximum they could get people to pay.[6]

Today, many executives see business ethics as being narrowly limited. They believe that when enterprises strive to maximize their profits while operating within the limits of the law, capitalism will ensure that the greatest good is produced for the most people. The proponent of free enterprise, economist Milton Friedman, expresses it this way: "So the question is, do corporation executives, provided they stay within the law, have responsibilities in their business activities other than to make as much money for their stockholders as possible? And my answer to that is, they do not."[7] Some executives go further. They feel justified in ignoring edicts they see as unfair to their enterprise, such as taxes or environmental and safety regulations they consider overly restrictive or beyond the right of government to impose. In his book, *The Ecology of Commerce*, businessman-author Paul Hawken quotes Harry Gray, a former chairman of United Technologies, as saying in 1983, "Such barriers as quotas, package and labeling requirements, local content laws, inspection procedures . . . inhibit world trade. We need conditions that are conductive to expanded trade. This means a worldwide business environment that's unfettered by government interference."[8] A number of people in business go still further in their philosophy. They see making money as "getting the better of the other guy." For some corporate directors, driving your competitor into the ground is the "name of the game."

All of these attitudes are troublesome. Although businesses theoretically have no desire to influence anything beyond their limited objectives, their impact on almost all human activities and the environment is enormous.

Eyes on the Gold, Forget the People

If you or I were to go out on the street and shoot somebody, we would be sentenced to life in prison or even be executed. When the officers of a corporation, in order to make a profit, foster activities that lead to the painful deaths of many thousands of people, they may be rewarded by handsome salaries, respect, and status. While some companies are being sued for damages, and some executives may be convicted for committing perjury, they are not likely to be treated the same way street criminals are for the deaths they perpetrate. This seems to be exactly the case, for example, with the managers of large American tobacco companies, their lobbyists, and their supporters in Congress. If these companies merely sold a dangerous product only to people who were aware that they could be harming themselves, it would be immoral. However, they do more. They promote the use of tobacco among people who do not use it and confuse people about its dangers to their own health and the health of others by withholding data and, until recently, repeatedly refuting the statistically demonstrated dangers involved.

Many of the promotional activities of tobacco companies are well known to Americans. We know less about what these companies do in developing nations where scarce money is desperately needed for public services and business investment—rather than for importing cigarettes.

Until 1986 foreign-made cigarettes were banned from Taiwan. All cigarettes sold there were made by a state-run monopoly which never advertised. The U.S. trade representative threatened to use trade sanctions against Taiwan if it did not open its market to foreign cigarettes, so it did. Hundreds of billboards glorifying smoking appeared throughout the island. Advertisements in magazines were directed toward youth and free cigarettes were handed out in discotheques. Within two years, smoking among high-school students increased from 19.5 percent to 32 percent, and percentages of women, very few of whom had previously smoked, skyrocketed. Similar practices with comparable results have been carried out by American tobacco companies in other countries such as Thailand.[9]

While I was teaching architecture in China, huge letters spelling "KENT" appeared on a local television tower. They could be seen for miles. Billboards advertising American cigarettes sprang up in prominent places around the city. Students told me that it was "cool" to smoke expensive American cigarettes. Should current trends continue, epidemiologists estimate that smoking will kill roughly 50 million Chinese who are now children. According to the United Nations' World Health Organization, in less than thirty years ten million people around the world will be dying every year of tobacco-related diseases.[10]

It would be interesting to read in the annual report of a tobacco company: "Last year was a good year, our 'operating income' was roughly $22,000 for every U.S. death that could statistically be attributed to our cigarettes.* Fortunately, we were not held responsible for

---

*"Operating income" can be considerably higher or lower than net income. For this example the following figures were used for calculations: According to Philip Morris's 1996 Annual Report, their retail share of the U.S. cigarette market was 47.8 percent and their operating income for domestic tobacco was $4,206 million. Figures from the 1995–96 *Statistical Abstract of the United States* indicate that cigarettes account for about 94.6 percent of tobacco use. *Cancer Facts and Figures—1997,* American Cancer Society, estimates that 419,000 U.S. deaths can be attributed to tobacco use each year. Using these figures, I arrived at an "operating income" of $22,000 for every death statistically attributed to smoking Philip Morris cigarettes. So one can say that for every cancer death, Philip Morris earns $22,000.

the medical costs of these deaths. They were paid by the people themselves or the government." (In the wake of the recent lawsuits against tobacco companies, however, this is changing.) Were these deaths worth it? Apparently the tobacco companies think so because they not only provide cigarettes but help smokers on their way by aggressive promotions and by fighting all efforts to restrict sales.

There is more to read in their annual reports, such as the words of Geoffrey C. Bible, CEO of the Philip Morris Companies. Besides describing increasing sales, profits, and market share, Bible writes, regarding attacks on his company: "Our U.S. tobacco company has faced similar threats before and has overcome them. . . . We are committing all the resources necessary to defend the company from new forms of litigation, making sure we have better firepower than our foes, no matter how formidable. . . . Beyond defending ourselves, we are turning the legal tables on some of those who attack us. We are going on the offensive to vindicate our civil rights."[11]

Philip Morris is not out to kill people; its purpose is to make a profit. However, the fact remains, according to statistics, it makes $22,000 for every person who dies in the United States as a result of smoking its product.

By investing in a company we condone its activities. We think that large, well-known companies that provide jobs and places for people to put their savings must be good. Nonprofit organizations such as universities and labor unions also help legitimize these companies by investing in them. Most of us do not look beyond the surface. Few investors look askance at companies that indirectly cause harm. It is the balance sheets that count. Fortunately, a small but increasing number of brokers, investors, and writers have developed an interest in what is called "ethical investing."[12] They are very concerned about how companies in which they invest behave. The world would be a much better place if we all were.

ETHICS IN POLITICS AND GOVERNMENT

We allow, and even expect, governments to do things we do not permit ourselves to do. War, which is murder unless it is in self-defense, is acceptable if it is "in the national interest." Aggression can always be presented as furthering this end. Hypocrisy, lying, deceit, and clandestine operations carried out against other nations are seen as normal, necessary activities of governments and of the people who run them.

Although we may complain, governments often get by with levels of irresponsibility that we do not accept in individuals: Nations too spend their assets (natural resources) as quickly as possible, go into debt, pollute themselves, and let the future take care of itself.

Most of the unstated real goals of a nation are not those of its most noble citizens nor even the higher aspirations of the common person. The national goals are more likely to represent the average or even baser ambitions of the public and business interests. These ambitions become honorable when attributed to the homeland. We like this, for we can let loose our primitive impulses, get what we want, and be proud of our country. As theologian Reinhold Niebuhr has written, patriotism "transmutes individual unselfishness into national egotism. . . . The unqualified character of this devotion is the very basis of the nation's power and of the freedom to use power without moral restraint."[13]

"IN THE NATIONAL INTEREST"

When our government says that it is acting "in the national interest," it is giving ethical support to what is really organized selfishness. Supporting such action becomes patriotism. Questioning such activity is a nonquestion; it is suicide for a politician and makes individuals unpopular and even hated by their peers.

Many people do not see higher principles above those of the authority of their church, their government, or some leader. When someone is critical of the United States government, many say, "love it or leave it." Some people even consider Lt. William Calley Jr., who was convicted of ordering the massacre of innocent civilians at My Lai during the Vietnam War, a heroic patriot. Yet these same people criticize Nazi Adolph Eichmann, who like Calley just "did his job" when he carried out his heinous duties as directed by the creator of the extermination camps, Heinrich Himmler.

Crimes committed for selfish personal reasons are paltry compared to the atrocities like these that have been committed in the name of God or fatherland. The consequences of such acts become more serious as the world becomes more complex and our future more tenuous. Kenneth Boulding, speaking of nations, warns,

> those moral qualities which make for internal harmony and may increase its survival from that point of view, may limit the society's capacity to deal with external enemies in the external environment. It is not surprising, therefore, that moral codes are frequently and almost universally ambivalent in this regard. They severely discourage murder, theft, cheating, lying and so on at home while encouraging these things abroad. . . . We now seem to be approaching the stage in the history of the human race when the union of the internal and the external ethic will have to be achieved; otherwise the external ethic will destroy us all.[14]

Power Corrupts

Another problem inherent anywhere power or wealth is concentrated is corruption. Like buzzards around a corpse, opportunists willing to do what it takes gather around activities they see as ways to increase their own wealth, power, or status. In an attempt to counteract this, governments have instituted rigid procedures, lengthy forms to fill out, reviews, and investigations. These consume a sizable part of a government's resources and talent, making it hard for it to perform its duties effectively and efficiently.

ETHICS ON A GLOBAL SCALE

Many people on our planet are hungry and homeless. Between 1988 and 1990 chronic malnutrition affected 33 percent of the people in Africa's developing nations, and the situation since then has only grown worse.[15] In 1970 there was only one country (Chad) that had a declining Gross Domestic Product (GDP) per capita, but ten years later thirty-five countries were in this category. To make matters worse, each year underdeveloped nations must pay roughly $60 billion to meet debt payments and to buy needed goods.[16]

Ervin Laszlo, advisor to the director general of the United Nations Educational, Scientific, and Cultural Organization (UNESCO) notes that not everyone is suffering. "By 1988 . . . the revenues of the seventeen largest industrial manufacturing companies (about $922 billion) equaled the revenues of fifty of the world's poorest countries, the home of 65 percent of the world's people."[17] As the world's rich nations are becoming richer and the poor ones poorer, in the United States, rich individuals are widening the economic gap between themselves and the poor.

This growing disparity appears to be acceptable to most of the people and governments of wealthy nations. Past efforts to help impoverished nations have often been frustrated by misdirected expenditures, mismanagement, and corruption. Efforts to help often produced few results beyond the further enrichment of dishonest individuals in these countries. Combine this problem with the selfishness and inability of people in the donor nations to apprehend the plight of the unfortunate and with the efforts of politicians in wealthy nations to please their constituents, and a powerful force is formed that resists helping. When conservative politicians, columnists, and radio talk show hosts put down and sometimes ridicule all government efforts to help the unfortunate at home and abroad, we are given welcome excuses not to help. We can keep our money without feeling guilty. Dishonest foreign leaders, our own lack of imagination, selfishness, and politicians seeking our support work together to keep a large part of humanity in a state of increasing misery.

Harm from Good Intentions

Well-intended efforts to help people that fail to address causes of problems can easily do more harm than good. We want to provide food and medicine to people who are starving, but without providing such people with means for controlling their numbers, this help will result in even greater misery in the future when even larger numbers of people will starve. In this case our efforts to behave ethically work against nature's laws where individual welfare is subordinated to that of the species and the planet's ecosystems. When these are damaged it will come back to haunt us. This is a dilemma we avoid discussing or even thinking about.

There is another problem related to the inequality between peoples that has frightening implications. The citizens of wealthy nations demand that their leaders continue to raise their standard of living—and they must do so simply to avoid unemployment and please business. The people of the world's poorer nations naturally feel that they deserve to have the same standard of living that North Americans and Europeans enjoy and they are working hard to catch up. This race will make it extremely difficult to contain global warming, however. How can we do this when there is nothing to discourage Americans from driving more miles every year, and when poor countries demand a higher standard of living for their people?

Bai Xianhong, a senior government scientist, explains China's position: "China must develop, and its people must enjoy a better life, but we can't make it without energy. . . . You can't say that for the sake of lowering carbon dioxide emissions, China shouldn't burn coal anymore."[18] China is now the fastest growing producer of greenhouse gases,[19] although it emits only about 13 percent per capita as much as North Americans do. Its potential for producing these gases is not just worrisome, it is terrifying.

We are caught between two opposing, unrelenting forces: the drive for a higher standard of living for all of the world's peoples and the environmental limits within which we must live. We could resolve this problem fairly, though with great difficulty, but those who can do something about it choose do nothing. It would be political suicide for leaders to tell their constituents that they must live modestly. Any journalist to suggest this would soon be left with only a small audience.

ETHICS NOW ESSENTIAL

In the past, by merely caring for ourselves and those around us, we functioned as a positive element in the process of evolution and the health of our planet. This is no longer the case. Because of the power we now possess, we have burst out of the constructive constraints of the evolutionary process and the earth's ecological system.

We are now holding the future of our species and of our planet in our hands, and not very carefully. General Omar Bradley is quoted as saying, "The world has achieved brilliance without wisdom. Ours is a world of nuclear giants and ethical infants." Ethics are not just a nicety, today they are essential. In chapters 13 and 14, I will offer some suggestions about what we can do to raise our behavior to a higher ethical level.

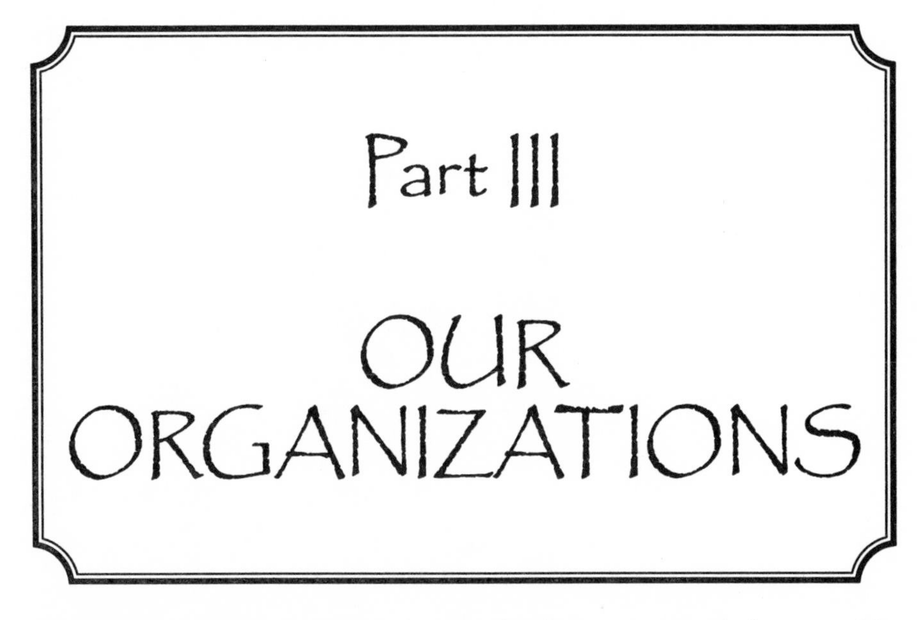

In human society, structured organizations are needed to enable people to do things they cannot do as individuals or informal groups. Although essential for a civilized society, organizations, in pursuing their interests, often come in conflict with each other and endanger the welfare of our species and the planet.

9

Leadership and Followership

In these difficult and dangerous times nations should have the most intelligent, informed, and competent leaders available, who should be doing their very best to deal with the problems before them. A quick check will show that this is hardly the case.

In the United States officeholders and candidates running for president and Congress have little time for contemplation and learning. Their days are filled with meetings and events. The problems of running a country are very different from those of winning a political campaign, but once they have won, winners cannot turn all of their energy to performing the duties of their office without great risk to their political futures. While in office in the early 1990s for example, incumbent senators felt compelled to raise an average of $2,000 a day for their 1996 (re-election) campaigns, and congressmen, $1,000 a day.[1]

Politicians must build and keep alliances. They must consider the political ramifications of every vote they cast as well as its consequences for their nation. Many of our most serious problems are simply avoided because they require politicians to take stands that could cost them votes in the next election. Raising taxes to obtain sorely needed revenues is an act to be avoided at all costs, as is terminating obsolete bureaucracies that provide jobs at great expense.

Considering the demands on their time and the type of thinking these demands engender, it is hard to imagine how elected officials can concern themselves with anything but the most pressing problems. Because so much of a government's energy is spent maintaining itself, little is left for it to do what it is supposed to do—managing problems effectively.

To appear effective while in office, a president must limit him- or herself to achieving a very few "major accomplishments" such as President Richard Nixon's opening of relations with China. This is not enough, especially today. Many serious problems confront us. None can be ignored. They must all be worked out or we will soon face serious consequences. This fact has not gotten through to us, though. The end purpose of politics still seems to be politics.

DOERS VERSUS THINKERS

Some people are curious and devote much time to learning about the world around them and to contemplating how conditions might be improved. We can call these people "thinkers." They recognize that complicated problems rarely have simple answers. While these people often lack leadership ability or the talent for getting things done, they generally have the best understanding of problems and the clearest ideas on how complex issues might be resolved. But seeing and dealing with complexity can render decision-making difficult and slow.

Other people concentrate on pursuing power, wealth, or fame and are often successful at gaining it. Consequently, the world is almost totally run by them. Their consuming interests in these areas, and the energy they expend on them, generally preclude them from being thinkers. Their curiosity is focused more on "how to do" than on "what should be done." They are "doers." Because they are more interested in doing than in learning, their knowledge in many areas is limited. Often they have little understanding of the dangers that await. Nevertheless, they "know" that their judgments, based largely on public and interest group pressure and expediency, are right. Many of these people fail to recognize their shortcomings. They need to make quick decisions, providing simplistic answers to complex questions. They give little thought to the distant future.

Herbert A. Simon, Nobel Prize-winner in economics, wrote about this type of person in his book *Administrative Behavior*:

> Administrative man recognizes that the world he perceives is a drastically simplified model of the buzzing, blooming confusion that constitutes the real world. He is content with this gross simplification because he believes that the real world is mostly empty—that most of the facts of the real world have no great relevance to any particular situation he is facing and that most significant chains of causes and consequences are short and simple. Hence he is content to leave out of account those aspects of reality—and that means *most* aspects—that appear irrelevant at a given time. He makes his choices using a simple picture of the situation that takes into account just a few of the factors that he regards as most relevant and crucial.[2]

Businesspeople and politicians are generally administrative in this sense. They are the people who run things, who get things done—or prevent them from getting done. Because they are successful and powerful, and because they think that what they are themselves good at is what is important, many of them view thinkers as ineffectual, irrelevant, irritating, and not worth listening to. When they look for advice, they generally turn to people who share their opinions and tell them what they want to hear, or to people who can tell them how to maintain or increase their power or wealth.

At times it *is* necessary to make quick decisions based on limited information. However, many decisions, such as those relating to the environment, education, poverty, and crime prevention, require clear understanding and time for careful consideration. Here doers, with their expedient solutions, can lead us into serious long-term problems.

Most doers are also realists. As young pragmatists, they see where power and opportunities lie and learn to work in the world as it is in order to get what they want. They understand each other and the rules by which they all play. Because these rules have been good to them, they want to keep things much as they are. Unfortunately, we are ever more dependent on such people to deal with very serious problems.

HOW LEADERS GET THERE

Do we select the people who govern us or are we selected to be governed by them? Democracy contains a bit of both, but it is closer to the latter. We get to vote, but our choice is limited. We get to choose from a list of ambitious individuals who generally have an appetite for power or a desire to be important. They have spent much of their adult lives building political support and maneuvering their way onto the ballot.

Political candidates are packaged and sold, and selling candidates is like selling anything: It all depends on making them appear to be something special. Gustave Le Bon saw prestige as the mainspring of all authority: "The special characteristic of prestige is to prevent us seeing things as they are and entirely to paralyze our judgment. Crowds always, and individuals as a rule, stand in need of ready-made opinions on all subjects. The popularity of these opinions is independent of the measure of truth or error they contain, and is solely regulated by their prestige."[3] Put simply, we select our leaders based on how familiar their names are, and politicians know it.

The selection process in totalitarian countries and those in a state of chaos is less subtle. Power is up for grabs by the person who is the most cunning or who can make the best deals with powerful interests. Dictators such as Stalin and Franco have demonstrated how far some people will go to gain personal power—inflicting untold suffering on millions of their fellow citizens in order to do so. It is amazing how helpless we are in preventing seemingly mad men driven by vanity and greed, such as Uganda's Idi Amin, from grabbing control of a nation and leading it into terrible trouble and barbarities. There are always individuals and groups who for their own benefit will help these tyrants do so.

People who desire and attain power are likely to have the traits I have just described. They spend much, if not most, of their time bolstering their power base and honing their political skills. People with opposite characteristics prefer knowledge to power, cooperation to competition, are future-oriented, and better see the complexities of difficult problems. These very qualities make it difficult for them to gain power or succeed in politics, however. They are not taken seriously and are often taken advantage of by unscrupulous doers. When they are in positions of power, these well-intended individuals have difficulty being effective leaders in the "real" world of sharks. President Woodrow Wilson, for example, had this problem.

Because of the people in it, politics is confrontational. Because it is confrontational it draws people who like confrontation, therefore it is confrontational—a positive feedback loop. It eliminates the kinds of people we desperately need today. They don't want to participate—and cannot. If they tried they would seem to be fools, or at least the media would make them appear to be.

"If some people act on the enlightened level and others on the egotistic level, in the short term the egotists win out and the enlightened people become marginalized," Ervin Laszlo states. "Conscious action even in light of the perceived whole, even at the level of nations and nation-states, would be far more advanced today if it were not for this situation. Leaders who begin to act more in the general human interest find themselves at a disadvantage in the short run. They often lose the game before they reach the point where their actions pay off."[4]

This mechanism affects the relationships between nations and cultures as well. Those whose values and beliefs favor material growth and aggression eventually dominate others, who must then join the carnage in order to survive.

Most superior individuals who have the qualities that should make them good leaders in our current complex world are not likely to run for office. They do not want to subject themselves to the hassles necessary to succeed in politics, are not prone to win the support of special interest groups, and are not likely to have wide voter appeal. Many do not want to spend their time and energy on political haggling, deal making, and the essential business of fund raising. Should a capable person actually be elected to office, his tenure would probably be a disaster. His efforts to cooperate and to solve problems objectively would be crushed in an environment where the rules of the game are very different.

We therefore have a mechanism that fills positions of power with people who can effectively deal with each other in their dog-eat-dog world, but who are poorly equipped to understand, and have little interest in, the complexity of many matters affecting our future. For some, campaigning and a hard-won victory over an opponent seems to override their interest in the job itself. Adlai Stevenson, an unsuccessful candidate for president, is reputed to have said that "by the time a man is elected president he is no longer worthy of the office." The late economist Kenneth Boulding, who was interested in the political selection process, held a similar opinion. "There is indeed a principle which I have called the 'dismal theorem of political science'—that most of the skills which lead to the rise to power [make people unfit] to exercise it."[5]

THE RESULTING DILEMMA

The people we must depend on to make important and often complex decisions to protect our future are some of the least able to understand the threats to it, or to register concern for it. In politics a limited viewpoint focused on current pressing issues and political survival is essential. To succeed, one must quickly recognize opportunities for personal advantage and react to them. Few people, including politicians, look beyond the parts to see the big picture. Since it is rare for outstanding human beings to reach positions of political power, we are often dependent on myopic and mediocre minds to make important decisions affecting the future of humanity and our planet. Actually, we have demonstrated that we do not want superior individuals to lead us. We no longer elect Adamses, Jeffersons, or Madisons. C. A. Gibb puts it succinctly: "the evidence suggests that every increment of intelligence means wiser government, but that the crowd prefers to be ill-governed by people it can understand."[6]

Those who gain power have a tendency to see nature as something to be exploited rather than as our sustenance. Money and other material things are more important and real to them. They appear unable to understand that we totally depend on the natural systems we seem intent on devastating. Their limited imaginations also make it hard for them to empathize with people living in circumstances different from their own. When public pressure mounts they feel compelled to take action, or more likely to appear to take action—and yet not disturb the status quo. Such public pressure led to the climate change conference in Rio de Janeiro which was "regarded by many as mere window dressing, as a pitiful attempt to show the world that there is a will to do something, although in fact there is not."[7]

With power having gone to their heads, some people in government believe that only they can be trusted to act wisely. For what they see as security reasons, they keep important information from the public. In such cases we are unable to act in our own interest on issues involving our safety, and some of the most knowledgeable people are kept ignorant and excluded from making decisions. For example, against the advice of some of the wisest scientific minds available and without public debate, politicians have built up huge accumulations of radioactive waste. They did not have a satisfactory plan to dispose of it, and still do not. They have exposed thousands of people to nuclear radiation without informing them. The United States and the former USSR continued making nuclear weapons long after they had built enough to destroy each other many times over.

Our well-being requires us to consciously constrain our instinctive drive to extend our place in nature. To do so, we need a clear picture of what is happening. Unfortunately, rather than recognizing current changes that could alter trends, some leaders look to the past for trends that they simply project into the future. They take examples of successful past expansion and project them into the future without understanding that there are limiting factors. Some nostalgic leaders seem to believe that we can recapture desirable qualities and ways of doing things from the past which actually are no longer attainable.

Humanity has worked hard to expand its ecological niche by fighting its enemies and increasing its food supply. However, eventually, limits that cannot be pushed back will be reached as population, consumption, and waste production increase. To magnify the problem, activities such as the destruction of topsoil, the contamination of air and water, and the dispersal of nondegradable toxins actually shrink the maximum possible size of humanity's ecological niche.

For many leaders, "success" means winning an election or surviving a crisis. In reality underlying problems may have grown worse and opportunities to correct them may have been lost. But the fact that these politicians got through this year is confirmation that their approach should be good for next year as well.

THE COMMONPLACE CITIZEN

With rare exceptions, leaders are competitive people who have a strong ambition to be on top. They need those who are led. Those who are led need leaders to create and manage an orderly, safe environment in which they can conduct their lives.

Those who are led, the public, can be viewed as individuals, each with her own special interests, and also as a homogeneous mass with a somewhat predictable set of needs, concerns, and demands. Not every individual shares the desires of the mass, however enough do to enable us to characterize it. (For example, not every American wants an automobile, but enough do to enable us to say that Americans want automobiles.)

This "public" exerts a powerful influence over what is done, and often it is not for the good. Spanish philosopher Jose Ortega y Gasset described the problem of the modern world as this: "The mass crushes beneath it everything that is different, everything that is excellent, individual, qualified and select. Anybody who is not like everybody, who does not think like everybody, runs the risk of being eliminated."[8]

Although their campaigns do not reflect this belief, American politicians often say "I have faith in the judgment of the American people." Ortega y Gasset is not so generous regarding the average citizen: "The mass is . . . those who demand nothing special of themselves, but for whom to live is to be every moment what they are, without imposing on themselves any effort towards perfection; mere buoys that float on the waves."[9]

While the "masses" ask little of themselves, they have unrealizable expectations of their leaders. Prominent among their demands are perpetual prosperity, security, comfort, and entertainment. They tend to give their allegiance to politicians who flatter their ignorance and apathy and promise to meet their unreasonable demands. As well intended as most people may be, their drives and goals are dangerous in this world today.

The public generally prefers charismatic leaders to good administrators. They want action! They want their problems solved. But they do not want to be bothered with knowing about difficulties or the possible negative side effects brought on by achieving their desires. As part of a crowd we are not accountable to anyone—we can be irresponsible and fickle. After viewing television broadcasts of atrocities committed in Somalia, we wanted our government to get involved. However, after seeing pictures of a murdered American being dragged through the streets of Mogadishu, we insisted that our government get out.

As I mentioned in chapter 5, some of the most catastrophic events in history have begun with the personal problems, idiosyncrasies, fanaticism, and psychoses of potentates. Mentally disturbed megalomaniacs such as Hitler, Stalin, Ayatollah Khomeini, and Jim Jones have had devastating effects on the people who followed them and often on many others.

Unfortunately, mutually supportive relationships are established between disturbed charismatic individuals who have a pathological need to be followed and frustrated, insecure, or paranoid people eager for easy answers to their problems, fears, and anxieties. Each party satisfies the warped needs of the other. In addition to the calamities that are generated in this way, many real needs such as ensuring peace or protecting the environment are ignored.

Many natural and social scientists and others know much that is wrong with the environment and much that can help us, but this knowledge finds few takers. People do not read, watch, or listen to things that disturb their beliefs or lie beyond their usual range of interests. Can one blame them? It is not pleasant to do so. Awareness, concern, and commitment to solving a problem begins with one or a few thoughtful individuals who lend it their support. But as the idea spreads, it increasingly comes up against disinterested, indifferent, unimaginative people who show little regard for what they do not directly experience, including future generations. Often this segment of society encompasses the majority of people and is very likely to include political and financial leaders.

IS DEMOCRACY'S WATCHDOG WATCHING?

We depend on journalists to report important happenings and keep us informed about human activities and the state of our planet. Ideally, these "watchdogs of democracy" provide us with the information we need in order to be good citizens. They are supposed to—and are actually guaranteed the freedom to—expose the bad behavior of government when it steps out of line. But do they? Can they do it as often as needed?

What newspapers and television actually do is shaped by many factors, including the interests of the public they serve, the desire of their owners to make a profit, and the personality types of the people who are attracted to media careers. The limitations of these people affect the content of the news we receive. Most journalists think in concrete terms, are fascinated by conflict, and are highly oriented to the present.

Unless subsidized, the media are obliged to support themselves by earning money, and often this is their primary purpose. People who buy stock in a media company (or any company, for that matter) generally do so because they think it will make more money than the stock of General Motors or Exxon or whatever company they choose *not* to buy stock in. To produce a profit, the media must provide readers with what interests them because advertisers gravitate to larger audiences. With few exceptions, journalists do not stay in business by trying to lead. They must follow. So what we get is plenty of news about food, celebrities, violence, human interest stories, and some conventional political reporting, but too little about serious matters requiring public participation, especially regarding the future. Newspapers and news programs that have failed to satisfy the public's wishes cater to small audiences or are no longer in business.

Because of the large amounts of money involved, the owners of newspapers and television networks tend to be corporations or very rich people. High officials and anchorpersons on network television news are also highly paid. This must give some bias to the information they present to us. Since the media depends on advertisers for financial support, it cannot help but also be influenced by the business community. The "watchdog" is not as free, objective, and critical as we would like it to be.

Journalists tend to reflect popular thought, particularly what conventionally educated people are thinking. These people are rarely interested in serious ideas or the more far-reaching aspects of issues relating to science, the environment, education, or crime. Many reporters take the easy way out by just quoting what people say, such as "this bill will never pass as is," or "we have the situation under control," rather than giving us background information and a clear idea of what is really happening. They leave us with unconnected fragments, endless unnecessary details of current conflicts, and little overall understanding of what is happening in our world.

Readers and viewers assess the importance of things largely by the space or time devoted to them. For example, from 1989 through 1991, television network news spent 67 minutes per month reporting on crime. A June 1993 ABC poll showed that 5 percent of those polled felt that crime was our most important problem. By the end of 1993 crime reporting on the networks had risen to 157 minutes a month. A February 1994 poll showed that crime had become the nation's most serious problem for 31 percent of Americans.[10] It went on to become the most important issue during the fall 1994 political campaign. While crime was becoming a major public concern, statistics showed that most forms of it were actually declining.

Thus, it is easy to see that the attention newspapers and television devote to different topics has little to do with their true magnitude or seriousness. Consider how little of the "news" we get is needed to be an informed citizen. It would be fascinating to see, spliced together in a single piece, one year's reporting and discussions of the rise and fall of stock prices from the *Wall Street Journal* and other publications. How many hours in all did intelligent minds in our country devote to reading this? Or to sports?

There are, of course, journalists who do describe and discuss issues and problems in depth, but their influence is limited and they too have a limited point of view. The people who follow them do so largely because the reporter reflects their own opinion—liberal, conservative, libertarian, Catholic, or oriented toward a profession such as medicine or an interest such as guns—to begin with. By limiting ourselves to the popular media and several other carefully selected sources, we have broadened our data base, but it is tilted in the direction of our interests and reinforces viewpoints we already hold.

The lack of real leadership at the point of information dissemination is particularly unfortunate. It is here that new facts and ideas can be introduced to the public, that a sense of importance should be established and balance maintained. Instead, following public interest, little attention is given to the world population problem or soil deterioration, but much is given to the batting average of some baseball player and the personal problems of the British royal family.

THE MYTH OF AN INFORMED CITIZENRY

For a democracy to work as it should there must be an informed citizenry that acts responsibly and ethically to further not just its own welfare but the overall good. However, the media want business, so they minimize presenting us with things that are uncomfortable.

Politicians and special interests can often achieve their ends by deceiving the public about what is right and what benefits them. In so doing they cheat the public out of exactly what a democracy is intended to do for them. Two examples demonstrate this point.

First, some years ago the billboard lobby, using paid advertising and interviews, successfully persuaded the citizens of Cincinnati to vote against a referendum to expand controls on billboards. Although one may wonder why billboards are good, the lobby was successful in persuading the electorate that they were desirable. Second, and even more clearly, a last-minute blitz of television commercials blaming Michael Dukakis for the polluted water of Boston Harbor helped swing the presidential election to George Bush in 1988. In reality Bush's environmental record was worse than Dukakis's and harbor cleanup had been stopped because President Reagan had cut off funding for it.

Deception works. It is a tool waiting for the unscrupulous to use to their advantage. It rewards perpetrators—confidence men, hucksters, and politicians—and it puts scrupulous candidates who avoid it at a disadvantage, while making it difficult for anyone seeking public office or furthering a cause to shun the practice. By wanting to believe, by fooling ourselves, and by openly regarding deception as part of the game, we give it respectability—and invite it.

Politicians and special interest groups want themselves and their concerns to be seen in the best possible light. Public relations professionals are ready to help. Bringing a candidate or an issue to the attention of the public is a worthy pursuit, but image building and "public education" rarely stop there. They hide dark sides and create sunny sides that do not exist. We voters, the ultimate power, are presented a "reality" that may be largely illusory. We can then easily be duped into voting against our own interests. As the tools and techniques for misleading and confusing become more powerful, and as issues grow in number and complexity, the mechanisms that are supposed to make a democracy work may become so misused and distorted that democracy ceases to exist, except in name.

There is another more ominous aspect to citizen participation. Many of us do not even bother to vote, or we vote without trying to inform ourselves about the issues and/or candidates. Once, when volunteering to drive elderly people to their polling place, I asked a woman which candidate she preferred. She said, with sincerity, that she did not yet know, but it would come to her when she was in the voting booth. She was very proud that she was doing her duty as a good citizen by voting.

WE ALL FOLLOW EACH OTHER

Political leaders are under enormous pressure to meet public demands. They must promise good times and avoid dealing with unpleasant subjects like fiscal responsibility, raising taxes, conserving resources, or a more equitable sharing of the world's wealth. We not only like politicians who promise good things, we want to believe them. Most of us tend to think that they understand what they talk about and do, but what they are doing is largely only what makes them popular with us, and what they know may only be what they need to know to win the next election.

Corporate executives know they are accountable to shareholders for maximizing profits, quality control, treating employees well, and protecting the environment. They know that the public buys what it wants—or can be made to want—not necessarily what it needs. Successful corporate leaders are also followers, i.e., they follow the desires of the consuming public and the stockholders. One way in which they do this is through "marginal differentiation." Automobile manufacturers and politicians in particular want their models and images to stand out, but only slightly. One year all cars will take on rounded shapes, but in slightly different ways, and all politicians will call for increased military spending, but by differing amounts.

Symbiotic relationships form between large numbers of uninformed people (who like comfort, ease, and conformity) and politicians, journalists, and businesspeople (who want power, influence, or

people, Lincoln and Roosevelt would have had no followers—and we would not remember them.[14]

We actually like inaction on the part of our leaders. Even when we recognize a serious problem such as the greenhouse effect, we may prefer to have our leaders ignore it because correcting it may require unwelcome changes in our habits. We want to have our desires satisfied now, and expecting tomorrow to take care of itself, we refuse to accept responsibility for our own future. Successful leaders know this and gain our support by encouraging our irresponsibility.

In spite of all this, we like to think that in some way we are in control. Actually, society and its leaders take no conscious control at all. We have no overall plan or goal. Events are guided by reactions to the unconscious drives of the public and its leaders, and to conditions around them. Procedures, rules, and programs are established as easy fixes when conditions become insufferable, or to provide the means for satisfying natural human urges. We let things happen, sometimes mending them, sometimes tolerating them without ever looking beyond the surface. Television produces violent programs to make a profit; a president loses an election because the economy goes bad; a dictator is supported because of chaotic conditions. All of these are reactions.

Government by the Lowest Common Denominator

In a democracy the majority and special interests set the agenda that politicians must address. In a typical campaign, each candidate must limit discussion to a few topics, generally aspects of material well-being, fear, security, and subjects of concern to special interest groups. The mechanics of democratic governments limit what they can do. They can, within departments, fund studies, make reports, and hold conferences on important matters. But ultimately, they are limited by what the political process and their citizens allow. Senator Al Gore observed, "Ironically, at this stage, the maximum that is politically feasible still falls short of the minimum that is truly effective."[15]

The problems governments face are often fuzzy and complex. It is hard to draw boundaries around them and find clear-cut solutions. The people involved, the issues, the political and economic consequences, public opinion, and hard-to-predict side effects all add to the confusion. Objective problem solving in government is nearly impossible. In politics it is difficult to present data accurately: Who says something, how well it is said, how closely it conforms to truisms and prejudices, and how often it is repeated have more impact on us than reason and the validity of the data itself. Where issues are dealt with on a public level, many must be simplified and stripped down so that they can be understood by a demanding public that refuses to inform itself. In the process, issues are often handled in a superficial way, ignoring many important implications. It is in this murky arena that we must work out our destiny. For example, the "three-strikes-and-you're-out" policy requiring life sentences for people convicted of three felonies is an appealing slogan meant to appeal to people fed up with crime.[16] It has little to do with reducing crime but much to do with gaining political support and increasing the cost of government. During the 1993 congressional battle of the budget, to protect their careers, members of Congress fought against increasing taxes and cutting benefits for their constituents and contributors. Years of work to build effective programs or conduct beneficial scientific research can be eroded or destroyed by politicians playing to the lowest common denominator and greed. The *Cincinnati Post* (March 18, 1985) quotes John Gilligan, former governor of Ohio, as saying, "The trouble with politicians is everyone is talking baby talk. To talk real issues would scare the hell out of their constituents."

Politicians win elections by telling people they can have everything without sacrifice. This not only fills public offices with people of dubious ethics, but it assures the public that responsibility and sacrifice are unnecessary. On winning, candidates are pressed to become the image they have projected themselves to be, and are forced to make efforts to fulfill some of the irresponsible promises they have made.

Another trick that wins votes is to blame opposing candidates for highly visible problems. To attract the support of white suburbanites, office seekers accuse their opponents of fostering government waste and being soft on crime. Hitler rallied support by blaming the Jews for many of Germany's troubles, and absolving the German people of culpability. As long as we see the "enemy" as being out there and not within, we feel good, but we cannot solve the problems that plague us.

Nations quickly point the finger at each other for precipitating war or damaging the environment. Developed nations blame the Third

World for deforestation, rampant population growth, and mismanagement. The developing countries point out that the per capita resource consumption and pollution created in industrial nations far exceeds their own. They claim that trade restrictions imposed by wealthy nations are forcing them to cut down their forests and produce raw materials for consumption so that they can pay their debts. Both are right about each other, and both need to change. But by blaming others and avoiding hardship for their constituents, leaders gain support.

If someone happens to take a responsible position, it is often politically advantageous for others to oppose it. This discourages elected officials from introducing views that counter popular opinion. The result is that many issues do not receive open debate but are settled by public emotion instead. This provides no incentive to get to the bottom of problems. Instead, it encourages dealing with symptoms in superficial ways.

When we don't bother to inform ourselves regarding all sides of an issue, but demand quick and easy fixes, we are asking politicians to deceive us. Consequently, they limit their discussions to what we want to hear and ennoble our shortcomings. The term "silent majority" glorifies apathy.

SERIOUS CONSEQUENCES

Politicians and the media place the economy and international hostilities at the top of their agenda, relegating the question of a viable future for our species to a spot low on the list of priorities. The economy and national security are important, but what about the very future of humanity? This issue is there, right along with others such as school prayer, gays in the military, the shape of conference tables, who's going to pitch for the Reds, and how a politician sees himself viewed by history.

We and our "leadership" are confronted by two forces. On the one hand there are human demands that cannot be ignored. No politician dare say, for example, that we must stop economic growth. On the other hand there are natural systems that are harder to understand but cannot be circumvented without serious consequences. We cannot continue to overfish the seas in perpetuity. Something must give, and in the end, it is nature that will not. Our method of governance, however, the "circle of followership," is a feedback loop that makes sure human demands are met—at least for the present. In this circle, nature has no advocate—until conditions become intolerable. Then it becomes very clear to us that nature's interest must be our interest as well.

Satisfying the public's desire for plentiful cheap energy helps keep politicians in office. However, it leads to rapid resource depletion and contributes to global warming, although these costs are not paid until later.

An uninformed, irresponsible public places demands on governments that can only be met by disregarding the future and mistreating the environment, other nations, and minorities. The leaders of wealthy nations dare not tell their constituents that they must lower their standard of living, nor can the leaders of poor nations tell their people that they cannot aspire to live like Americans. As long as people and nations can invent ways to benefit by blaming others, many pressing problems cannot and will not be solved.

The "circle of followership" leaves us in a dangerous position today. We are like a little dog that chases its tail while a big dog is pursuing it. Too many politicians prefer to react to issues as they arise and to solve crises when they occur rather than initiate a plan of action and pursue publicly stated goals. If this is true, it leaves us adrift in a world of danger with no real leadership. As an architect who enjoys solving problems, I find this hard to understand.

We can do better, though, as we shall see in part four.

10

The Workings of Government

Governments do for nations what brains do for people: They manage. They perceive what takes place within and around themselves, store data, and direct action. Nevertheless, like all forms of life, governments' first and overriding priority is self-preservation. People in government, elected and hired officials as well as usurpers, give their main attention to preserving and advancing their own positions. Dictators use armies, secret police, censorship, purges, and torture to do this. In democracies, elected officials focus their activities on winning the next election. Bureaucrats kowtow to their superiors and comport themselves in ways that protect them from criticism. Those who succeed survive. They run governments. In all fairness, many conscientious individuals perform their duties well and provide many useful services as effectively as they can under these trying conditions.

The most important thing about a government is who actually controls it. This affects its character, what it does, and how it does it. Control is determined by the structure and working mechanisms of the particular government—and even more importantly, by human nature, as discussed in chapter 9.

ECONOMIC SYSTEMS

One of the key functions of a government is to adopt and uphold a specific economic system. The citizenry may share everything as needed, all may be the property of a leader who dispenses goods and demands services as he sees fit, or the arrangement may be something considerably more complex. Let us look at the two systems that have had the most impact on our planet and our species in modern times.

Communism

In theory, Soviet-Maoist style communism professes to stand up for the needs of working people. In practice, the reality is more ambiguous. Absolute power is claimed by communist parties. Leaders and officials see to it that their positions are protected and that they are well treated. The benefits of party membership attract opportunists and people desiring power for its own sake. Most thinkers, creative people, and idealists are repelled by Soviet-Maoist communism's intellectual constrictions, its hypocrisy (members really *aren't* treated equally), and its time-consuming political meetings.

While many of the original communists were idealistic, as they left government and were replaced, personal well-being and political maneuvering became the primary concerns of party members; citizen welfare became secondary. Privilege, bribery, and corruption are intrinsic to such a system. People who are aware of and concerned about the environment, along with the less fortunate and those promoting civil rights, have no power at all and dare not speak out. Central planning has proven to be a poor way to meet people's needs and allocate goods and resources. Government funds are funneled into armaments and to the secret police for the purpose of keeping the leadership in power.

In the communist system, government interference and the lack of rewards for one's efforts undermine people's ambition, which significantly lowers production. I had a chance to view this system from the inside to a degree while teaching at a university in China in 1992. The control party officials held crushed much of the enthusiasm professional people had for their tasks.

Capitalism

Capitalism effectively responds to market demands and motivates efficiency. Because capitalism has recently been so productive and communism has failed, capitalism is touted as a quasireligion, especially by people who are more interested in profits and extending property rights than in protecting personal rights and freedom. At this time, the achievements of capitalism are blinding us to its weaknesses.

Laissez faire capitalism has many remarkable self-regulating mechanisms and incentives that stimulate invention and hard work. On the other hand, as we shall see in chapter 12, it exacerbates a number of serious problems, including the exploitation of workers and natural resources. By having placed controls on capitalism, we are spared some of its most negative consequences. Unfortunately, there is little money to be made and little thanks are given for doing what is needed to protect our planet or to promote peace. Generally it is just the other way around.

Ideologies Distract from Real Problems

Political and economic systems affect people's relationship to nature and to each other. Pure capitalists and Marxists concentrate on the relationships among people and completely disregard our relationship with the earth. During the Cold War moderated forms of these systems competed so strenuously with each other that governments concentrating on this conflict had to largely ignore many problems, including those of the environment. During normal times, under "controlled capitalism," such concerns can be addressed. However, with the aid of powerful lobbies and public apathy, they can easily be cast aside. Presidents Reagan and Bush, by believing in minimum regulation of the marketplace, reduced funding for the unfortunate, workers' safety, education, the arts, and environmental protection. Marxism, which claimed to be the benefactor of working people, has left a terrible legacy of social, cultural, environmental, and economic damage.

It is hard to give up or change systems that work well in ways that seem important to us. Marxism, as it was practiced in the USSR, provided power, security, and privileges to those who controlled it. Capitalism provides wealth and security to those who succeed in making money. The earth, the poor, and other politically and economically weak entities that are bypassed or harmed by these systems have little say in changing them.

Taxes and Subsidies

Within capitalistic systems, money acts as a catalyst. It stimulates activity and affects human behavior. Money can be used, loaned, moved, or stored. Governments can utilize this tool in the form of tax structures, subsidies, and controls to channel social energies in desired directions. On the other hand, tax deductions can motivate businesses to spend lavishly on entertainment, yachts, country club memberships, and sales meetings in remote luxurious locations. Shortsightedness by the government and greed often cause tax incentives and subsidies to be misused.

If a resource is inexpensive, its use is accelerated and people become wasteful. To illustrate, some years ago, the price of natural gas was artificially held down by the U.S. government. This caused two things to happen. People were inspired to use gas with little concern for waste. Profits for producers became so low that exploration and the development of new wells stopped. A shortage developed.

The Worldwatch Institute (an independent, nonprofit environmental research organization based in Washington, D.C.) maintains that taken together governments around the world direct at least $500 billion a year to activities that hurt the environment, for example, building dams and draining wetlands, and this figure does not take into account other harmful subsidies. Since few countries have tried to assess the actual amount of the subsidies they provide, the total may be far greater.[1] Why do we spend so much money doing things that hurt us when the money is desperately needed for other purposes?

The fact is we really do not seem to mind being fooled. We do not like to pay taxes and we complain that too much is spent on this and that. We pressure our representatives to cut back. However, we ignore most subsidies that aggravate the problems we face. The parties who receive them, however, care very much about them and see that politicians who uphold them stay in office. These parties do not believe that we would all be better off without most of the payments or tax benefits that now exist, so such practices continue.

In the United States oil depletion allowances* and low taxes on petroleum encourage indiscriminate driving. After the Arab boycott of the 1970s, the government required manufacturers to produce more fuel-efficient cars. We now drive more efficient cars and petroleum costs have even dropped. However, because fuel costs were reduced, our cities spread out as suburbs proliferated, so we drive much more, negating the savings. With the oil crisis fading from memory, gas-guzzling, four-wheel drive recreational vehicles exempt from efficiency requirements have become fashionable. Our unquenchable thirst for imported fuel feeds our national debt and makes us dependent on often unfriendly nations for our life-blood, oil. Nations with higher taxes on petroleum also suffer these problems, but to a lesser extent.

Had we significantly raised our taxes on gasoline years ago, we would now be less threatened by future boycotts and possible wars over oil reserves. Problems caused by the burning of hydrocarbons and suburban sprawl would be smaller. However, such rational taxes are unpopular with voters and opposed by powerful interest groups.

Excusing the producers and consumers of products and services from responsibility for the harm that they cause removes any restraint on their use and passes the cost of damage on to others. When people who produce or use electricity are not held responsible for harm resulting from sulfur dioxide and carbon dioxide emissions resulting from the burning of fossil fuels in power plants, they have little incentive to conserve. People who smoke may give cancer to others, but this does not deter them from smoking because they do not have to pay the ensuing hospital bills. A study made in 1992 at the University of California in San Francisco estimated that the cost to the state of a package of cigarettes was $3.43, mostly in lost wages and higher health care costs.[2] (The average cost of a pack of cigarettes was probably less than $2.00 at that time.)

---

*A means for reducing income taxes by amortizing the oil extracted from an oil field. The amount amortized can exceed the original cost of the oil field, which is very advantageous for the producer because it reduces his cost of production.

In 1993, Americans paid a hidden subsidy of approximately $2.25 for every gallon of gasoline they bought. It came from their property taxes, and can be broken down as follows: The cost of roads, police patrolling, courts, and debt service on highway bonds equalled $0.29. Curbing air pollution and related health costs came to $0.09. Subsidies to downtown employers when their employees used free parking amounted to $0.17. (A business could deduct $155 a month for giving employees free parking but only $60 for mass transit, so it was more beneficial to them to encourage their employees to drive.) The cost of protecting Persian Gulf oil supplies (assuming that one-half of this was for the benefit of motorists) was $0.20. Another $1.50 covered the expense of automobile insurance, injuries and deaths resulting from auto accidents, as well as the time lost from work as a result of those accidents and the many hours businesses and individuals lost to rush hour traffic. The total of $2.25 did not include the cost of street lighting, handling storm water as a result of the extra runoff from paved areas, and the disposal of used vehicles, tires, and batteries.[3]

Real estate taxes affect how land is used and how buildings are constructed and maintained. Taxes based on the cost or market value of a building discourage people from erecting high quality, attractive buildings and give property owners reason to let structures deteriorate. High taxes on structures and low taxes on land encourage speculators to keep prime parcels of land vacant, or make them into unsightly parking lots (rather than space-conserving garages), with the hope that someday the land will rise in value and produce a profit. This method of taxing promotes the conversion of productive farmland into low-density suburban sprawl and forces people to consume huge amounts of petroleum in order to move about in it. At one time the federal government even provided funds to communities so that water and sewer lines could extend into the countryside, which accelerated the growth of suburbia.

Taxes on oil wells, mining, and logging would encourage the conservation of resources for future generations, provide funds needed to correct environmental damage, and pay medical costs incurred by the extraction and use of resources. Such taxes would be compensation for damage, and payment to society for the privilege of extracting resources. Instead, interest groups have successfully pressured the U.S. government into providing subsidies and tax breaks to drillers, miners, and loggers. This has kept the price of goods artificially low in relation to the cost of labor. We therefore waste cheap materials and energy in order to save the expensive cost of labor. This is especially ironic at a time when many people are unemployed.

We produce far more goods than people really need. We squander huge amounts of paper and plastic in packaging and advertising. We ship by truck and travel by automobile and aircraft—the most fuel-consuming means of transportation. We reshape and extend our cities so that we must travel great distances in private automobiles to work, shop, or do almost anything. However, we do save on labor.

The logical, equitable idea of a "Green Tax" to compensate for damage done to others and to the environment has been around for a long time, but of course, it is unacceptable. It was first proposed by the English economist Nicholas Pigou, who published the *Economics of Welfare* in 1920. Pigou argued that if producers did not bear the complete cost of production, which should include environmental damage and resulting sickness, the competitive marketplace would not work as it should.[4] This faulty model is what we must now contend with.

CONGENITAL WEAKNESSES

National governments are incapable of meeting all of the important needs of the modern world. The confrontational individuals who normally lead governments (as discussed in chapter 9) take adversarial stands on issues, making agreement slow and difficult. This prevents governments from addressing many of the serious problems society faces. Confrontation perpetuates itself, which makes this situation hard to change.

So much energy goes into these political conflicts and into maintaining authority, and so many commitments are made in doing so, that often governments have little energy and latitude for dealing with a wide array of serious problems. Power struggles, pleasing an ignorant constituency, and satisfying special interests pervert much of the legislation that is created and result in programs, tax rules, subsidies, and entitlements that, taken together, are chaotic and sometimes counterproductive.

As I mentioned in chapter 6, it is politically impossible for an American president or Congress to take on more than a few major legislative projects during a political term. However, considering the real world beyond politics (there is one), and the number and seriousness of the problems we now face, it is disastrous not to deal with them. Put bluntly, our government is incapable of doing what it must. We face each problem individually, and we do not modify the structure and mechanisms that prevent us from doing better.

Many problems are very complex and have interested parties lined up waiting to defend their turf when the public attempts to address them. "Horse trading" and back-scratching often mold such solutions, making thoughtfulness and fairness nearly irrelevant. Overlooked and unforeseen consequences may create even worse problems in the future.

To keep up with increasing world demands for food, we have devised a variety of methods for increasing food production. Unfortunately, these methods require up to fifty times as much fossil fuel energy as the energy the food itself contains. In the United States, due to erosion, we have been losing five tons of topsoil for each ton of wheat produced, reducing future productivity of the farmland.[5] Governments and agronomists are pleased with current levels of production, but they give little thought to what people will do when fossil fuels become scarce and topsoils approach depletion.

Another problem we are handling poorly is uncontrolled population growth. Growth is the worst in developing countries where there are good reasons for large families. Poor families in India, for example, want at least four sons. Some may die in infancy. One or two should go to the city and send money back home, while the others help at home and try to expand the family holdings. Unfortunately, since one is just as likely to have daughters, this leads to very large families. In a society that cannot afford government support for the aged, having children to care for them in their old age is the only means by which the elderly become secure. Acting responsibly and limiting family size is dangerous in such a society. What is a necessity for the family is disastrous for the nation—making it even more impossible for the nation to care for its aging citizens.

Developed nations that try to help their less fortunate members have a very different problem, which Richard Dawkins describes:

> If a husband and wife have more children than they can feed, the state, which means the rest of the population, simply steps in and keeps the surplus children alive and healthy. There is, in fact, nothing to stop a couple with no material resources at all having and rearing precisely as many children as the woman can physically bear. But the welfare state is a very unnatural thing. In nature, parents who have more children than they can support do not have many grandchildren, consequently their genes are not passed on to future generations.
>
> Contraception is sometimes attacked as "unnatural." . . . The trouble is so is the welfare state. . . . But you cannot have an unnatural welfare state, unless you also have an unnatural birth control, otherwise the end result will be misery even greater than that which obtains in nature. . . . Any altruistic system is inherently unstable, because it is open to abuse by selfish individuals, ready to exploit it.[6]

The cost of supporting teenage pregnancy is high and is a rapidly expanding burden on society. Former U.S. Surgeon General Joycelyn Elders testified before Congress that in 1992 "The government spent $34 billion on Aid to Families with Dependent Children, Medicaid, and food stamps for families begun by adolescents."[7] Considering that many of the most fit people do not even reproduce their numbers, with the best of intentions, we may be throwing the process of evolution into reverse.

Sometimes it seems impossible to find a good solution to a problem. Reality can push us from different directions, there being good arguments on both sides of an issue. In 1992 the government of the Gujarat State in India opposed the World Bank's proposal to cut funds for what would be the environmentally and socially disastrous Naramada Dam, which would control flooding and provide irrigation water for a number of years. The government felt that they could not deny short-term relief to people dying of hunger and thirst at the risk of precipitating future catastrophe.

Expediency Rules

Expediency is a favorite tool of politicians, who, in some developed nations, provide weapons technology, nuclear know-how, and nuclear materials to neutral nations to gain influence over them—or just to make money. Such deals make it easier for unstable leaders and terrorists to gain access to dangerous weaponry. Because expediency and "realpolitik" dictate giving respect to nations who possess nuclear weapons, we provide nations with an incentive to go nuclear. Expediency then dictates that we give nations such as North Korea political "carrots," for example, light water nuclear power plants (which do not produce material usable for making nuclear weapons), to forego doing so. Such incentives make it all the more attractive for developing nations to achieve nuclear capability.

The Problem of Time Lag

No one may be aware of a problem until it has already caused considerable damage, as was the case with depletion of the ozone layer above the earth. We discourage people from bringing us such essential information by calling them alarmists, radicals, troublemakers, or by ignoring or isolating them. Even so, a small, usually powerless part of the public may take notice. As a result of their persistence, in time, mainstream journalists may pick up on the problem and bring it to the larger public and arouse their concern. This opens up a period of conflict. People take sides and special interest groups may join in to dispel apprehension and diminish concern. Debate ensues, largely confined to name-calling, labeling, and shouting slogans. Public concern may swell, but then abates over time. One issue that illustrates this cycle is the rising cost of health insurance. American anxiety about this topic eventually forced politicians to take up the issue in 1994. After an unsuccessful attempt to resolve the problem, other issues, such as welfare reform, captured public attention and the problem was never solved.

Eventually public outcry or pressure from special interest groups becomes so great that politicians can no longer ignore the problem. Then the political process begins: Referrals are made to committees; hearings take place; studies (perhaps taking years) are commissioned; battles break out between factions and between Congress and the administration; bargaining, deals, and turf wars occur among government agencies; bills are proposed; more politicking takes place; and finally there is a vote. This process can take years. Sometimes public concern languishes along the way and much to the relief of elected officials the process is quietly dropped. Other times steps are actually taken, but they are often weakened by loopholes favoring special interests. Once enacted, it takes time before legislation becomes law and bureaucracies can incorporate it into their operations. Even though the government has now become engaged, albeit inadequately, the problem may be far from being solved, as I will demonstrate.

Let's consider those climate-changing emissions from motor vehicles, for example. Although almost everyone recognizes the problem, there is little effort to actually do anything about it. All solutions are "unacceptable." They all involve sacrificing something we dearly love or paying more for mobility. If we insist on keeping our automobiles and suburban way of life, we will have to switch to a source of energy that is probably considerably more expensive than petroleum.

Alcohol, coal-based fuels, hydrogen, or electricity have been mentioned as possible substitutes. Unfortunately, there are technological difficulties to overcome before those become viable alternatives. Crops needed to produce fuels such as alcohol would require vast areas of farmland in a world where demands for food will soon exceed production. But let's assume that electric propulsion became practical. What would conversion involve?

New vehicle models would have to be designed, tested, and refined. Automobile plants and hundreds of suppliers would have to retool or close and start anew. Jobs would be lost, people would have to relocate and communities would die. This would all take time. But there is more. We would have to replace the current infrastructure that supports the internal combustion engine, including oil fields, tankers, refineries, gas stations, garages, and mechanics. It would be hard to sell electric cars when there are no mechanics or recharging stations for them. Likewise, it would be hard to set up a network of recharging stations and electric car mechanics when there are few electric cars. How will we get people and the infrastructure to convert while petroleum is still available? And what would workers, factories, and others in remote locations face due to the more costly electric energy for transportation?

This conversion would take decades, but we are now in the first part of the process—gaining an awareness of the problem and of our own inaction. As nature writer Bill McKibben notes, our past intentions can hardly be called serious. "Although the United States promised at the 1992 Rio conference that we wouldn't be emitting any more carbon dioxide in the year 2000 than we did in 1990, we've done virtually nothing to meet our promise, and like almost every other developed nation on earth we will miss the goal, and probably by more than 10 percent."[8]

The Cost of Inconsistency

When government leadership changes, there can be major changes in policy and personnel. These changes can alter laws and their enforcement, funding, and programs in progress. Where long-term consistency is essential, such as in promoting energy efficiency or controlling water pollution, a change of direction or a cut in funding can be devastating. One research project was stopped in its tracks when President Reagan appointed as managers of the Environmental Protection Agency individuals hostile to environmental protection. Years of research were rendered useless because the collection of data that had to span a five-year period was suddenly cut short.

Restoring committed leadership and funding after something like this cannot fully mend the damage that has been done. Dedicated researchers and bureaucrats know that similar occurrences can happen again. Due to periodic changes in leadership, there is no assurance that long-range government programs can be completed. Inconsistency is demoralizing and it discourages talented individuals from working for the government. This works to the advantage of those hostile to government action of any kind because it increases government's difficulties in getting things done.

Warped Incentives

Businesses must operate efficiently, meet new challenges, and reorganize themselves as necessary in order to survive and prosper. On the other hand, governments often stick to policies and practices that no longer make sense. They do not have the stockholder incentive to adapt that businesses have. Instead, they are subjected to powerful forces that oppose necessary change.

Hugh Troy, a famous practical joker, demonstrated our willingness to accept red tape and nonsensical ways of doing things while he was at an army camp during World War II. Frustrated by the huge number of forms he had to fill out every day and forward to the Pentagon, he came upon the following idea. He devised and mimeographed a form called the Flypaper Report. It showed the floor plan of a mess hall he had to report on, and carefully located and numbered the ten flypaper ribbons that hung from its ceiling. Every day he noted how many flies were caught on each ribbon and included the report along with the others he sent to the Pentagon.

About a week later two of his fellow officers, who apparently had been asked to submit their own copies of the reports, asked him if he was "catching any hell from Washington" for not sending in his Flypaper Reports. He kindly gave his colleagues copies of his report and explained how they were to be filled out. They took them and passed them on to other officers, all of whom then faithfully filled them out and included them with their other daily reports. As far as Troy knew, this may have then become a standard procedure for the U.S. Army.[9] Senseless routines and mindless emulation come easily to us.

Once established, programs, regulations, and procedures must be followed—no matter how nonsensical they are. They may even introduce unintended, possibly unnoticed, side effects that do much, or even more, harm than the good they were intended to achieve. For example, laws intended to provide security for people working as maids or gardeners, because they require complicated bookkeeping, have caused families to hire people illegally or not at all.

Politicians may lose constituents by introducing changes to correct such problems. Bureaucratic turfs may be threatened. Individuals and interest groups that profit by working within the framework of existing laws or programs fear that change may deprive them of their privileges. A reduction of international tensions represents a threat to investors, employees of the defense industry, and others whose livelihood depends on the continuation of hostilities. Personal success and survival in government depends on avoiding unpopular action. Therefore, innovation, clever ideas, and dealing with the distant future are avoided.

Damaging Feedback Loops

Damaging positive feedback loops creep into political and economic systems. Instead of restoring balance as negative systems do, they accelerate destabilization, increasing their own power. The use of the automobile forms a positive feedback loop. The more it is used, the more necessary it becomes to have one and drive it, which again furthers its use. This also increases the size, determination, and strength of groups that pressure government to protect the automobile's place in society and even to subsidize it.

As our environment degenerates, individuals and organizations that further its degradation become more powerful. As oil companies deplete reserves and as lumber companies cut forests, they earn money that enables them to work faster and lobby politicians more effectively. The cruel Myanmar (Burmese) dictatorship encourages loggers to cut Burma's forests at a rapid rate. This cutting, which seriously damages land belonging to the Burmese people, provides the government with the funds for buying weapons needed to retain their power and thus allow the loggers to continue to do their work.

Economic competition among nations rewards those that exploit their resources most aggressively and protect their environment and people the least. By not playing by these rules, a nation is less competitive, falls behind, is open to exploitation, and can lose control over its own destiny. In the end, only the biggest producers and the most ruthless exploiters can succeed in many areas of activity—unless even-handed protective measures are applied across national boundaries. It is not to a nation's short-term economic advantage to apply such controls unilaterally.

The prestige given to nations that have achieved wealth and power leads others to emulate them. Exploiting nature and people to achieve economic success has become both a habit and a necessity. It cannot be continued indefinitely without disastrous results, yet it cannot be stopped without economic, political, and social trauma. The dimensions of this predicament will increase as competition for world markets and scarce resources intensifies, and as a growing world population has greater expectations and more powerful technologies at its disposal.

FORESIGHT AND VULNERABILITY

The worst disaster at a nuclear power plant occurred in 1986 at Chernobyl in the Ukraine. Fallout killed or disrupted the lives of thousands of people and made thousands of square miles of the surrounding land too dangerous to farm. Did this teach us anything?

Years after the disaster, problems are still widespread. Julian Borger described the situation as it stood in 1994: Reactor number four, which had partly melted down in the original catastrophe, was poorly encased in a rapidly deteriorating concrete shell. Two other reactors of the same design, which had had improvements made to their electronic control systems, were becoming more dangerous by the day because their structural components were deteriorating. Every visiting foreign expert was "terrified" by what he saw. U.S. Deputy Energy Secretary William White was reported to have told journalists, "We would like to see it closed down immediately—today." Borger noted, "One diplomat estimates that 1,000 technicians left Ukrainian nuclear-power plants in 1993 for Russia, where they can earn eight times as much. Many less-skilled personnel have been left in charge of outdated, understaffed, and inherently unstable reactors."[10] When society ceases to function well and there are no good choices, as is now the case in the Ukraine, people become desperate, care little about the future, and are willing to take frightful risks. Hence, Chernobyl stays on line.

No one predicted exactly what happened at Chernobyl, but experts should have foreseen that a terrible accident was likely to happen. Warnings were probably made, but were ignored by authorities. With hindsight the 1994 warnings about Chernobyl should be taken even more seriously, but leaders do not—or cannot—respond.

The former Soviet states are now burdened with an immense amount of radioactive and toxic wastes, and with manufacturing plants and an electric power system that demands nuclear energy in order to avoid economic collapse and social chaos. What might happen to nuclear power plants that use fuel and produce waste that is costly to store and remains deadly for many millennia during periods of economic strife, war, or chaos?

Just to mention a small part of the problem, thirteen naval reactors and at least 17,000 corroding containers of nuclear wastes are lying at the bottom of the Kara Sea north of Siberia. Bellona, a Norwegian environmental group, estimates that in the Barents region near Norway and Finland, it would take $128 billion just to clean up radioactive wastes, which includes junked and active nuclear warheads and radioactive materials in deteriorating containers.[11]

Russia and some of its former territories are beset by formidable economic and political problems that threaten their stability. They cannot afford to take care of their most mundane needs. How can they afford to clean up their poisonous mess or stop it from growing? What will happen if current conditions deteriorate further and individuals are bribed to sell dangerous materials to other nations or terrorists?

In the United States we note the myopia and irresponsibility of past USSR leaders, but fail to take our own shortcomings seriously. Our efforts to clean up dangerous substances are far from adequate. We continue to produce nuclear wastes even though we lack an acceptable plan for storing them safely for the many thousands of years they must be stored. What will happen if our descendants can no longer afford the high costs of maintaining storage facilities? To avoid pandemics, they will have to maintain laboratories for monitoring diseases and developing new antibiotics, although we are unwilling to adequately fund these now.[12] Why aren't we more concerned about this?

Our ability to supply food, water, and energy to distant locations; remove waste; and move people, goods, and information have enabled huge metropolitan areas to develop. Modern public health measures permit people to live safely at high population densities. Recent developments in information handling enable modern governments, businesses, and transportation systems to do things they never could before. Our lifestyle has changed and we have become dependent on these fragile and sophisticated technologies and systems. We think little about our increasing dependency on them and our vulnerability to their disruption.

My wife, Angela, studied at the Academy of Music and Theater in Mannheim, Germany, during World War II. During this time she was bombed out three times, lost most of her possessions, and ended up living in ruins and basements. Having seen photographs of the almost total devastation caused by the bombardment, and having visited Mannheim several years afterward, I was curious how people could survive in the rubble that was left. Angela explained. Horse or ox drawn wagons would bring water from the nearby countryside into the city. They parked at specific locations where people could fetch water with buckets. Unlike our sprawling cities, Mannheim was dense and compact. People could get about by foot or bicycle when street-cars did not work. Most food was raised locally, often by farmers whom people knew personally. Farmers and many city dwellers knew how to keep many foods such as cabbage, potatoes, and turnips edible over extended periods of time. Few people had central heating systems that needed electricity in order to work. Space heaters could burn almost anything when there was no coal in the basement. People were not dependent on elevators to reach their home or workplace.

What was difficult for people in the bombed cities of World War II would be catastrophic for us today. We have gradually made ourselves highly susceptible to many kinds of catastrophes. What would we do if there were no electricity to run the fans in our furnaces, or if fuels were not available? Water lines would freeze and sewage would accumulate. The explosion of one nuclear device high in the atmosphere could destroy most of the silicon micro chips, rendering our computers, cars, trucks, telephones, radios, heating systems, businesses, and possibly even our government useless.

What if we could not get food or water into our cities and waste out? Boston cannot survive without food from distant locations and San Diego would be dry in a few days without water from Nevada and northern California. How would we control the inevitable epidemics, not to mention new diseases and existing ones that have developed immunity to antibiotics? Think about it.

An example of such lack of foresight includes an Oklahoma office building with large, inoperable windows designed by an architectural firm for which I once worked. The air-conditioning system broke down during the final stage of construction. After two hours the building became so hot inside that it had to be vacated. Consider also the 1977 power outage in New York state. At the time, some friends of mine were staying in New York City. The elevators in their hotel could not operate and the corridors and stairwells were inadequately lit. My friends stayed in their room for two days. They were lucky; they had windows that could be opened and they were not on one of the upper floors that required pumps for the water supply and sanitation facilities to work.

We have become sitting ducks for terrorists and crazies who are increasing in numbers and daring. Each ghastly act is followed by an increasing array of burdensome protective measures, but little thought is given to the form terrorism may take in the future, or to alleviating its causes. Forecaster Marvin J. Centon warns, "The next 15 years may well be the age of superterrorism, when they gain access to weapons of mass destruction and show a new willingness to use them. Tomorrow's most dangerous terrorists will be motivated not by political ideology, but by fierce ethnic and religious hatreds. Their goal will not be political control, but the utter destruction of their chosen enemies."[13]

Our Growing Dependency

The world moves into an increasingly precarious future, with people giving little thought or guidance to it. We now depend on reliable, low cost, but disappearing energy and other resources to maintain not just our lifestyles, but our very lives on this highly populated planet. Our greatest dangers may not as yet even have been imagined.

L. F. Ivanhoe, geologist and geophysicist, predicts that world demand for petroleum will likely catch up with supplies around the year 2010, causing sharp increases in retail prices. He notes, "The peak global oil finding year was 1962. Since then, the global discovery rate has dropped sharply in all regions."[14]

The reasons he gives for governments ignoring this situation are interesting—and all too plausible: "All government petroleum ministries have an inherent interest in announcing the 'good news' of large national hydrocarbon reserves, inasmuch as large reserves are useful for national political prestige and in negotiations for OPEC production quotas, World Bank loans and grants, etc."[15] In making their projections, they use total reserves of oil and gas rather than limiting them to petroleum that can be economically extracted. International estimates of world petroleum reserves are based on these misleading figures.

Every day we become more dependent on intricate systems that need highly trained people to maintain them at a time when large numbers of young people do not finish high school. While we are making ourselves ever more vulnerable to disaster, the possibilities for such catastrophes are increasing rapidly.

Inevitable wars, economic downturns, natural calamities, and a growing number of other problems will cause nations to direct resources away from preparedness for contingencies and long-term concerns. Many nations are approaching a state of increasing poverty and chaos and affluent nations are finding it more difficult to maintain their deteriorating physical and social structures.

The only thing that gets people and governments to think and prepare for disaster is disaster. This is dangerous. As time passes, the potential damage a disaster can cause increases, our dependence on essential supplies becomes more tenuous, and the tools available to terrorists become more varied. Someday they may include materials such as nuclear weapons, plutonium, and a more widespread use of biological weaponry.

A World Divided by Fences

Astronauts looking down on our planet see a world of snow, oceans, green land, mountains, and deserts. They do not see the national boundaries that partition our globe and are so important to us. What they do see are the effects of our work: denuded landscapes, human-made deserts, seas streaked with eroded soil issuing from deltas, and at night, fires from burning rain forests. It would be wise for us to see the world as the astronauts do—without its artificial boundaries. But for us, boundaries are far more important than natural features or the ominous damage we are inflicting on our planet. Instead of joining together to confront common problems, we prefer to fight each other, and by so doing, cause further damage.

Within well-defined sovereign nations, there may be unrest and violence at times, yet a sense of order prevails. Laws, and police to enforce them, maintain harmony. People insist on it. Between nations, however, there is neither effective law nor policing. People insist on this as well. We fail to notice that technology, business, and culture have created a new community that extends around the world. Now people living on opposite sides of the planet may interact with each other more than people living two hundred miles apart did two hundred years ago.

Although business and culture have become successful at reaching across borders, in most matters the absolute sovereignty of nation-states persists. Having been bypassed in many areas, nations tenaciously hold on to their function of protecting their citizens and waging war. With other interests crossing boundaries, it is ironic that we are so intent on defending states that are rapidly losing many of their original purposes. This may be due to the fact that national leaders are people who like to be on top, so they normally resist any global order that limits their sovereignty.

There is another way to view this problem. In evolution, there is movement toward ever higher forms of organization. Elementary particles came together to form atomic nuclei. Atoms combined to produce molecules, which eventually brought complex living creatures and the societies to which they belong into being. Humans organized themselves into increasingly intricate social groups eventually leading to modern nation-states. However, progress beyond this point has been resisted by world leaders and ultranationalists in spite of the great need for it.

Given the challenges we face, and knowing that we like to think of ourselves as intelligent, it is incredible that no person or body has been given the authority to responsibly coordinate human activities and oversee the protection of our planet. Nobody has even been asked to try to do this to even a small degree, except in a few very limited areas. In the United States we have laws and institutions that protect human rights, but on a planetary scale no such laws exist. Leaders are only accountable to their own nation, which in some cases means only themselves. As was the case with Idi Amin, under the guise of national sovereignty they are allowed to do with their subjects as they wish.

So far, we have been unwilling or unable to insist on more responsible behavior by ourselves and our leaders in our much changed world. We consider those who fought to free and then unify our original thirteen states great patriots, yet we often consider "unpatriotic" citizens who want to extend much-needed unification to the world.

PROBLEMS OF DEMOCRACY

The type of government a nation has affects the way it responds to the needs of its citizens. An autocratic or oligarchic government would be ideal under an all-wise, benevolent leadership. The government would be run well and justly administered and people would not be required to spend large amounts of their time observing it and following social/political issues. Sadly, most autocrats are not wise and benevolent, and the chances of a tyrant assuming control are very good.

As we saw in the last chapter, the reality of how our democracies work falls far short of the dream, and we all know it. We do not, and cannot, do what is needed to have our democratic governments operate at an ideal level. Because the alternatives to democratic rule are grim, we must sustain democracies as best we can. Wallowing in patriotism and flying flags while we ignore our shortcomings cannot sustain democracy.

Democracy has served us well in the United States—at least until now. However, almost everything has changed. The challenges we face are greater, and the consequences of our responses, or lack of them, are critical.

Self-Serving Groups

To be good citizens in a democracy, we should theoretically keep up with every issue of importance. As we have noted, this would require that journalists report all significant events, which is impossible. Due to this lack, democratic nations are left with citizens who are inadequately informed on important issues. Rather than attempting to learn something about everything (and failing to do so), some groups limit their concerns and concentrate them on specific issues, thereby learning a lot about a little. This gives them power in these areas, where the average citizen is weak.

Special interest groups, the public, and our planet are all affected by government decisions. Some special interest groups are altruistic, and pressure governments to protect the planet, the weak, and the general welfare. These groups are usually supported by donated funds and donated time, which cannot be recovered. Other special interest groups are more self-serving. Their time and money are investments from which they expect to reap rewards. They can afford to make substantial contributions to political candidates who further their interests. Political successes by altruistic groups are often accompanied by exhaustion, whereas the self-serving earn profits, part of which they reinvest into lobbying. This imbalance forms a positive feedback loop favoring selfish, narrow interests. Once such interests have been established, they gather strength and embed themselves in the system.

Powerful politicians are more likely to attend the meetings of the chamber of commerce, veterans' organizations, unions, and other well financed groups than those of politically weak altruistic groups. International corporations and large domestic ones have an additional influence on politicians, who court the jobs and income the companies would bring to their countries. These businesses can go elsewhere if their demands are not met.

Historically, the enactment of legislation on issues where the concerns of special interest groups conflict with public welfare follows a predictable format. An issue presents itself. Altruistically concerned people push for a solution. Special interests, using techniques they have developed over decades, resist in every way they can. They contest corroborating evidence and present conflicting results of their own "investigations." They call for more time-consuming studies. They claim that required standards cannot be achieved, that costs to consumers will increase, that businesses will fold, and that jobs will be lost. And—they make political donations. And they very often win.* In spite of this, we do and must go through the same procedure over and over again, considering the credence of every claim, no matter how ill-founded.

In the past, self-interest and special interest groups have fought

*A good example is the well documented struggle of Vernon MacKensie, former Assistant Surgeon General and Director of the National Center for Air Pollution Control, U.S. Dept. of Health, Education, & Welfare. MacKensie wanted the U.S. government to become involved in controlling air pollution, but was resisted in every way by industry and bureaucracy. See Shawn Bernstein, *The Rise of Air Pollution Control as a National Political Issue: A Study of Issue Development,* Ph.D. diss., Columbia University, spring 1982.

social security, employee safety regulations, food and drug controls, and a wide variety of environmental protection measures. Today the majority of Americans, including many interest groups, recognize these government functions as being essential, yet such groups continue to fight most new measures that are introduced to protect the public and the environment.

When the media report on conflicts between self-interest and altruistic groups, they present both groups as having equally high-minded reasons for their positions. It is easier and less costly for them to simply report the claims of each side and appear fair than to thoroughly investigate the matter at hand and objectively report what they learn (for example, that some of the claims made are completely without merit). This leaves the public confused. Politicians playing it safe listen to this confused public. When issues get hot, public opinion and the financial support of donors become more important to politicians than facts, which fall by the wayside. Politicians then produce legislation that gives the public the impression it is getting what it needs, but also leaves loopholes that meet the needs of special interests.

Lobbyists attending a congressional hearing can now relay messages about a stand taken by a key congressman to their offices with cellular phones. Their offices immediately alert their supporters by fax and Internet so that they can have a barrage of messages waiting for the legislator when he returns from his meeting. Electronics are helping to reduce meaningful debate and increase the power of special interest groups.

The Trouble with People as Citizens

A democracy is grounded in its citizenry. Its constitution and its traditions are essential, but without the participation of its people, these become hollow shells.

We all notice people who are more interested in order and security than in freedom of speech and tolerance, and those to whom property rights are more important than human rights. Some people are more interested in trivia and entertainment than their government, and an increasing number have developed a way of life or doing business that depends on government payments. Both conservatives and liberals complain about different aspects of all this. I believe that both are right. In addition, many powerful, well-financed interest groups narrowly focus their concern on their own welfare, totally ignoring the overall good.

We All Have Our Own Agenda

Small, single-minded interest groups can gain power by supporting each other. Because their concerns are so limited they care little about the agenda of other groups—as long as they themselves get what they want. The Republican party has gained control of Congress largely by satisfying demands of divergent interest groups. The specific agendas of groups like the gun lobby, Christian fundamentalists, and wealthy people wishing to lower their taxes, among others, make up a significant part of the party program. This cooperation gives single-minded interests what they want. It produces a total agenda that most people do not want, and that works against the common good and the future. It also makes it difficult for democracies to approach problems holistically, fairly, and wisely.

Totalitarianism is unacceptable. Democracies, as they are now structured, have difficulty dealing with their own rapidly increasing internal pressures, as well as with the demands of a world that is increasingly complex and precarious. As the problems we face increase in number and intensity, approaching these pressures merely through gut reaction or expediency will not work. The inevitability of disaster does not seem to occur to us or arouse our concern.

Democracies can be pulled in many different directions. Some people have an insatiable desire for riches. They push for a minimum of government controls and taxation. When they succeed, the rich get richer, the poor get poorer, and the environment and public welfare deteriorate. When there are rules but people and officials are dishonest, a widespread system of bribery and corruption develops, accompanied by human misery and environmental abuse. Or, a system of increasingly cumbersome order can prevail. People are basically honest and the government tries to right all wrongs. Bureaucracy and legalism grow like mold, resulting in a society of people afraid to take risks, making it exceedingly difficult to do anything that is not routine.

Seemingly oblivious to the dangers that loom, we go on, pursuing our personal interests, limiting ourselves to tinkering with the details of government, and relying on patches to get us through the next few years. There is no serious movement to study how we can combine the values we cherish with the needs of the new reality that is descending on us. Unless we prepare ourselves, we will either be unable to meet the challenges of a changed world, or we will lose our freedom.

WHY DON'T WE DO BETTER?

Today's governments, democracies and others, are fossils of the past. They simply were not designed to deal with today's problems. But the situation can be even worse. Psychiatrist Robert Jay Lifton points out that sometimes governments "just act crazy." One need only look at recent history in Germany, Uganda, Libya, Iraq, Cambodia, and Rwanda to see this.

With all the failures of modern governments, little creative thinking is given to making them work significantly better. Instead, we make many small adjustments that are acceptable to a lethargic public and special interest groups. Sometimes these changes are helpful, but at other times they merely give us the feeling that we are doing something.

Governing a nation, even in quiet times, is not easy. Einstein was once reported to have been asked, "Why is it that when the mind of man has stretched so far as to discover the structure of the atom, we have been unable to devise the political means to keep the atom from destroying us?" It is said that he answered, "That is simple, my friend. It is because politics is more difficult than physics."[16] Successful scientists are usually intelligent and knowledgeable, whereas almost anybody can go into politics.

11

Organized Violence

History is full of horror stories of killings and torture. Many of us thought modern nations had moved beyond that, but then came World War II and its numerous brutalities. Especially disturbing was that many of the worst atrocities were committed by two of the world's supposedly most advanced, educated countries, Japan and Germany. Since then violence has continued in many diverse places. It is unpleasant but important to understand what we are capable of doing as groups and nations in a state of frenzy.

Examples can be horrifying, as the following incident, described by *New York Times* correspondent Nicholas Kristof shows. (Kristof wrote this after he came across secret Communist party documents that attest to large-scale cannibalism in the Guangxi region of southern China during the Cultural Revolution.) "The first person to strip meat from the body of one school principal was the former girlfriend of the man's son; she wanted to show she had no sympathy for him and was just as 'red' as anybody else. At some high schools, students butchered and roasted their teachers and principals in the school courtyard and feasted on the meat to celebrate triumph over 'counterrevolutionaries.' "[1]

Ruth Leger Sivard records that between 1950 and 1996 wars caused nearly 22 million deaths.[2] Whether that is comparatively few or many, it is far too many for creatures who like to think of themselves as civilized. (The only acceptable figure would be zero.) There are, however, many reasons why reasonably intelligent, moral creatures turn to violence. Anthony Storr, noted English psychiatrist, comments on one of these: "It is clear that human beings possess a marked hereditary predisposition toward aggressive behavior which they share with other animals and which serves a number of positive functions. An animal has to be able to compete for whatever resources of food are available. Sexual selection is ensured by competition for mates. . . . Animals which live in groups tend to establish hierarchies which reduce conflict between individuals."[3] We humans have not been able to extend our groups into one large world group. So, although the world's people interact with each other, they have not established adequate hierarchies to reduce inevitable conflicts.

Even though the entire populace on both sides of a war may be embroiled in it, the number of people directly involved in starting or condoning it can be relatively small. The actual instigators of the aggression (generally those who have striven for and achieved political or financial power) are usually more bellicose, competitive, and greedy than the population as a whole. They may be people who lack talent in the arts, science, scholarship, technology, business, constructive statesmanship, etc., who find an outlet for their energy and gain importance by creating conflict and hostility. If Hitler as a young and aspiring artist had been admitted to art school, World War II might may never have occurred. His aspirations would have found an outlet in painting and he would not have satisfied his need for recognition and admiration in brutal subjugation. Society is so constructed that a few individuals can stir up and lead a normally peaceful population into war.

At considerable personal sacrifice, ordinarily nonviolent people will zealously join their fellow citizens to fight another nation or religious group. Belonging to the group, demonstrating their loyalty to it, and receiving its approval is important to them. When doing so they will proudly commit acts, including murder, that they would consider immoral in other situations.

As Storr has pointed out, rivalry and hostility played constructive roles in animal and early human societies. Within a group, they instituted a pecking order and stabilized relationships among individuals. Between groups, they established territories which reduced the likelihood of conflict, and by keeping population densities low, kept game animals from being overhunted. With the dreadful weapons we possess today, things are different. Within nations, hostility can produce murderous street warfare, between nations it can lead to annihilation—possibly of all life on earth.

We grow up in a world with a certain amount of violence, some clearly provoked and some resulting from misunderstandings or mistakes. We have a natural tendency to respond to violence in kind. It takes an unusual person to resist this tendency. Our impulse is to react to a wrong with another wrong, rather than to try and defuse the hostility. Revenge is not just a natural human feeling, but a deeply felt tradition in some cultures and social groups. Violence thus begets violence, which easily starts an escalating cycle that is hard to break.

For example, World War I was followed by the Versailles Treaty, which required excessive reparations to be paid by Germany. That, plus political instability and worldwide depression, led to Hitler's ascendance, setting the stage for World War II. The Holocaust provided impetus for the creation of a Jewish homeland, and European Jews were granted part of Palestine. This in turn has led to festering hostilities in the Middle East and cycles of revengeful terrorism and military retaliation which are difficult to bring to a close. Even so, World War II was different—and unusual. The Allies made a reasonable peace with Germany, Italy, and Japan, opening the way to a peaceful future between winners and losers. Too few have learned from this outstanding lesson.

If one of two adversaries chooses force, the second must either respond with force or capitulate. It is hard for a party to unilaterally choose peace and survive. In fact, it can only do so when its adversary can be overcome by reason, goodwill, moral persuasion, or international sanctions. Recall that the British did relinquish control over India, but it took Mohandas Gandhi many years using the power of "peaceful resistance" to convince them to do so.

When peoples' lives are wretched, they jump at simple answers. They fall prey to demagogues, become eager to blame scapegoats, and are more willing to use force. Bad times, instability, fear, hate, chauvinism, and peer pressure work together to enable a small number of individuals to seize political power and create hostilities. When this happens, others are eager to loose their most brutal urges and jump in. During the Great Depression, a period of extreme hardship in Germany, a few fanatical Nazis were able to rouse other Germans; grab power; start a war that drew in most of Europe, North Africa, North America, and part of Asia; and murder millions of innocent people.

In war we are willing to have our governments spend money in a grand way, ignoring the great waste of resources and environmental damage involved. Peace is different. Here we are miserly. During the Cold War we were willing to pay for submarines, tanks, and bombs, but we are now stingy when it comes to spending the "peace dividend" or to helping keep a friendly, peaceful Russia from possibly descending into a state of chaos that could be dangerous to the whole world. In 1993 the world spent more than forty times as much to prepare for war as it spent to promote peace.[4] We wage first-class wars, but are only willing to pay for fourth-class peace.

THE GLORY OF WAR

Many nations and cultures have idealized war and aggression. There are monuments, songs, paintings, stories, movies, and operas that celebrate it. Military leaders such as Alexander the Great, Caesar, Napoleon, and Patton are venerated.

Until recent times, many Chinese, Islamic, and Western weapons and armor were works of art. In eighteenth-century Europe there was nothing more exciting than a military parade with its horses, cannons, and soldiers in colorful uniforms. This glorification of war continued though the beginning of World War I, when people marched off to war with much hoopla and enthusiasm, and into the present with some people collecting uniforms, weapons, and other memorabilia.

The foolishness of this attitude is demonstrated by Julian Grenfell, an English poet. In 1914, Grenfell wrote to his mother, "I've never been so fit or nearly so happy in my life before: I adore the fighting and the continual interest which compensates for every disadvantage. . . . I adore war. It is like a big picnic without the objectlessness of a picnic."[5] Seven months later, he was wounded and died.

Historians have traditionally elevated the most brutal despots to positions of great importance in human history. Journalists who report current events go even further in bestowing importance on conflict and violence—it is the major part of their stock in trade. Feeding us with conflict becomes a financially rewarding positive feedback loop. We are drawn by it. Consequently journalists, television companies, movie producers, and toy and video game makers have perfected means for presenting violence to us in a variety of attractive ways, encouraging us to enjoy it all the more. We all get caught up in the chase, where even the good guy must resort to violence in order to restore justice.

An aura of importance and glamour is given to violence and other forms of strife. Constructive events and real contributions to society are inadequately reported or ignored. This affects to the very core the perceptions, concerns, and values of every one of us individually and society as a whole.

In many cultures, Latin countries and street gangs for example, young people, women as well as men, feel compelled to appear fearless. Machismo and appearing "manly" are cultural trademarks for some young men. Thus, they adopt a variety of behaviors in order to maintain this appearance. Violence is one of them. Cooperating, nurturing ("female") traits have a hard time standing up against competitive, dominant ("male") traits. With today's stress on sexual equality, violence is becoming unisex. Movies, television, and computer games are filled with women committing violent acts.

We cannot say that the Germans, the Japanese, the Serbs, or the Hutus are bad people. Our basic makeup is the same as theirs. Violence exists in us as it does in them. Many people, when they hear of gross injustice or torture, imagine exacting revenge on the perpetrators. Fortunately, civilizing forces and our own sense of decency constrain such urges. Until we stop pointing our fingers at others, recognize the potential for violence in ourselves, and address it, we will continue to perpetrate acts of brutality in the guise of doing good.

THE SACREDNESS OF SOVEREIGNTY

Almost all living organisms do things to improve their condition. Plants turn their leaves to the sunlight; protozoa move toward food by beating their flagella. Humans have more options. As their increasing numbers puts pressure on the environment, they can (1) work harder, (2) invent ways to better exploit their environment, (3) move, or (4) expand their territory. The first soon reaches limits, the second leads to invention, and the last two to exploration, migration, and eventually conflict with other groups of humans. The last option, expansion, seems to have become the favorite choice. In fact, many people like expansion so well that when they have met their practical needs, they keep pushing for more, encouraging aggression. In modern society, it is the last group, the aggressors, who have gained power and dominate all others.

Although the most important reason that led to the formation of sovereign nations, defense, has disappeared, nations still retain their traditional boundaries. Nuclear, biological, and chemical weapons can be delivered to almost any spot on the planet, leaving once secure civilians in the most powerful nations defenseless. While antiquated boundaries fail to defend, these artificial divisions of the world create serious problems for the environment and for people today. By uniting some people and excluding others, for instance, states, or groups of states, separate their citizenry from people who live outside their borders. Thus situations of "we versus them" are created. We judge other nations and peoples from our own limited viewpoint, applying different ethical standards when dealing with people from other countries or alliances.

Not long ago, there was deep concern for American veterans of the Vietnam War who were exposed to Agent Orange. Simultaneously, we totally ignored similar problems of the people we claimed we were fighting for, the Vietnamese. Their situation was a nonquestion. American lives are worthy of our concern, others are not. It does not seem to strike us that Billy the Kid, Charles Manson, and Al Capone were Americans, and that Jesus, Mother Teresa, and Beethoven were not.

We strip-mine, pollute, deplete topsoil, and cover our land with visual ugliness. Our government often defends the rights of landowners to do so. However, should any nation take one square foot of our land, we would go to war to defend it.

Leaders have often increased their own power and importance by exploiting or arousing paranoia in their followers. This is not difficult to do. A penchant for forming barriers and harboring hostilities lies deep in our psyche. Followers of one belief (communism, Christianity, Islam) may feel compelled to convert others to their belief—by force, if necessary. Communication between hostile political and religious groups is usually poor. Separation and hostility come easily, whereas integration is difficult. A world divided into camps of zealots, ready to convert or defend, as ours is, is a dangerous place indeed. Few national leaders have personality types that would allow them to surrender their sovereign power in order to affiliate themselves with other countries for the benefit of world order. This is demonstrated by the difficulties that supporters of the European Union have in gaining cooperation among national leaders.

THE "SECURITY DILEMMA" AND POLARIZATION

As long as the world is divided into sovereign nations, we will have what political scientist John Herz calls the "security dilemma":

> The "security dilemma" besets, above all, those units which, in their respective historical setting, are the highest ones, that is, not subordinate to any higher authority. Since, for their protection and even their survival, they cannot rely on any higher authority, they are necessarily thrown back on their own devices; and since they cannot be sure of the intentions of competing units, they must be prepared for "the worst." Hence they must have means of defense. But preparing for defense may arouse the suspicions of others, who in turn will engage in such preparation. A vicious cycle will arise—of suspicion and counter-suspicion, competition for power, armament races, ultimately war.[6]

Andrew Bard Schmookler sees the problem similarly but describes it in a different way: "The irony is that successful defense against a power-maximizing aggressor requires a society to become more like the society that threatens it. Power can be stopped only by power, . . . no one is free to choose peace, but anyone can impose upon all the necessity for power."[7]

When violence emerges, which is often, everything else is pushed back or forgotten. Values, ideas, and projects that have been painstakingly developed over long periods of time, succumb. Instead of spending our money on medical research meant to save lives, or on the preservation of art masterpieces of the past, we spend it to kill thousands of people and wipe out cities containing irreplaceable works of art in a single air raid. Instead of working to preserve ten acres of woods in our community, we spray many thousands of acres of land with defoliants to expose an enemy, killing almost all life and permanently changing an ecology.

Political leaders are inclined and pressured to think in the short term. They are caught up in the "security dilemma," finding quick, seemingly simple solutions to threats—which are unending. Leaders seem to give little serious thought to long-range solutions that would eliminate perpetual crises.

Modern times are beset with another problem which John Herz calls "penetrability." The primary purpose of nation-states was defense. Today long-range ballistic missiles fitted with nuclear warheads can penetrate the defenses of even the most powerful nations. Our progress has not been in civility, but in developing the tools to enable us to commit ever greater barbarities.

When we arm ourselves, we don't seriously consider how this will affect our neighbors. Inevitably it results in reactions directed back at us. Our decisions to arm are based on perceptions of what is happening around us and on internal politics. We fail to put ourselves in our neighbors' place, and consider how we would act should they start building up their weaponry. Our love of competition only speeds up arms races. We enjoy winning. Once an arms race or war starts, peoples' attitudes toward war becomes irrational, their view of it changes. Thinking that could have led to reasonable conflict resolution is replaced by fanaticism or something approaching it. This escalating cycle is a positive feedback loop.

When communications and travel were slow, weapons unsophisticated, and environmental damage not so extensive, there wasn't the need for international cooperation that there is today. Although the most powerful nations are now largely defenseless against terrible weapons and terrorism, we still continue to threaten, fight, and resist international law and its enforcement. We seem to be unable to comprehend reality, or to sensibly weigh the risks we take by tenaciously holding on to cherished sovereignties.

Political scientist Robert C. North criticizes our adversary method for solving many serious problems.[8] It turns problem solving into disputes with winners and losers, rather than constructive, objective deliberations. Perhaps this can be partly explained by the fact that lawyers compose a significant, powerful, well paid, and respected element of American society. This gives undue influence to legalistic ways of thinking which focus on details, tactics, and confrontation at the expense of substance and physical reality, such as the interaction of natural systems.

The global sovereignty issue can be seen on the microlevel when individuals who disagree try to resolve their differences: In the United States, the lawyers' "ethical code" will not allow a divorcing couple to go to one lawyer to peacefully work out an agreeable settlement. Each person must have his or her own lawyer, who turns the case into a "we versus them" situation. This means there will be a winner and a loser. Too often, only the lawyers win.

Lawyers perform an important service for society: They see that order is maintained. But in the United States (unlike Sweden and Japan, where there is far less litigation), our legal system has grown extremely complex, and we quickly turn to the courts to solve disputes. Our legal system has grown out of proportion, overstepped reasonable bounds, and become a positive feedback loop generating ever more need for itself. The presence of lawyers in our society has grown so powerful that their adversarial ways have spilled over into areas where they can only do harm, cause unnecessary expense, and inhibit new ideas.

This adversarial system thrives on differences of opinion, which easily become confrontations, leading to polarization. Polarization encourages us to resist hearing our adversary, deters us from recognizing the legitimate concerns of those who differ with us, and hinders our finding mutually beneficial, creative solutions. It can cause us to fight change where change is needed, or insist on change where it can cause harm to others and possibly ourselves. Confrontation introduces bad feelings and resentments, making progress toward mutually constructive resolutions difficult. It leaves many people feeling insecure and anxious. Losers may be left despondent or with a desire for revenge.

Hostility, even without confrontation, can produce damaging

results over a long period of time. During the Cold War, the USSR and the United States ignored many pressing problems while they diverted huge amounts of money and talent to producing weapons and arming seemingly tractable nations. They left a legacy of toxic chemical and nuclear waste, pollution, antiquated factories, and unsafe nuclear power that we now cannot afford to deal with effectively or safely. Competition between the major powers led indirectly to the proliferation of nuclear weapons and the potential for producing them among other nations. It is unclear exactly what this will lead to, but the possibilities—such as international terrorism, nuclear war to solve regional disputes, and the sale of nuclear technology to meet economic needs—are chilling.

The consequences of violence and threatened violence on a national scale reach beyond the brutalities of the moment. They leave a legacy that future generations must live with.

12

Business and Its Sidekick

Except in police states, business surpasses government as an influence over our lives. It affects how we interact with each other and with the planet. It even influences the way we think. It uses both the carrot (financial rewards) if you play along, and the stick (financial loss and social censure) if you do not. While it needs and asks for limited government controls, it has a far greater influence on government than government has on it.

THE PROFIT INCENTIVE

My father greatly enjoyed first selling, and later producing, leather for shoe and garment factories. Enjoyment of the task drives much that is done in business. Profit is another motivation, at times the only one. Business can be conducted without enjoyment, but not for long without profit. So when these come in conflict, profit usually wins.

Unencumbered free enterprise fosters efficiency, which is good. On the other hand, without regulation, it encourages the exploitation of natural and human resources and furthers pollution. The lowest cost producer prevails. Spending as little as possible for labor, worker protection, and pollution control gives a manufacturer an advantage over others.

Most of the time, the need to maximize profit determines what is done. The purpose of a business operating at top efficiency is not to produce bicycles, but rather to produce profit. If a bicycle factory can make more money with its assets and know-how selling insurance, in theory it will do so.

It can make economic sense, for example, to exhaust a natural resource rapidly. Gaylord Nelson, former senator and councillor of the Wilderness Society, was asked during a National Press Club talk[1] to explain why the Japanese do not reduce whaling when they know that at present rates of harvest whales could become extinct in about ten years. He explained that if one can make a 15 percent return on capital by bringing about extinction but only a 10 percent return by maintaining the world's whale population at a stable level, it makes short-term economic sense to do the former. Whalers would make a lot of money fast which they could then invest elsewhere. By doing so, the whalers would come out ahead, but the whales and the rest of humanity would be losers. This same principle applies to clear-cutting forests, overgrazing mountain slopes, and other ecologically harmful practices.

The market—supply and demand—allegedly determines what is done. But in the real world, it is not that simple. Businesspeople have learned how to tamper with demand by increasing it. Advertising and planned obsolescence are their tools. Advertising induces people to purchase goods and services they might not have otherwise purchased. Planned obsolescence increases replacement rates either by producing goods with a limited life, or by making them trendy so that one must keep replacing them periodically to stay in fashion.

Maximizing profits is good for business efficiency and benefits speculative stockholders but it can come at a high cost to people, communities, the environment, and ultimately business itself. We do not ask, "Is this what we really want?" However, businesses do ask, "Does this increase our profits now?" We have willingly become pawns moved about in the game of financial manipulation and by our own desire to have "name brands" at the lowest possible prices—which has helped close many locally owned businesses. Basing business decisions solely on profit creates a closed social-economic system, where human values and the welfare of the planet are left out.

Competition drives nations to promote economic expansion and new technologies, or lose their power and prestige. It drives nations to pollute and exploit their own and other nations' natural resources. It can create hostilities. By not joining this game, a nation falls behind, opens itself to exploitation, and loses control over its own destiny. In the end, in many areas, only the biggest, most powerful, and most aggressive succeed—all in the hallowed names of free trade and free enterprise.

Greed and ruthless competition set the stage for escalating confrontations and hostilities. Unfair apportionment of wealth and power was one of the sparks to hostilities in Northern Ireland, Nigeria, Yugoslavia, and Lebanon. Economic cruelties caused by the Versailles Treaty led to the takeover of Germany by the Nazis.

This perpetual diversion of human energy and talent distracts humanity from maintaining peace and solving its long-range problems. Many reasonable people who want to improve our condition become discouraged by this and by the insignificant effect of their efforts.

OUR LOVE AFFAIR WITH GROWTH AND BIGNESS

In chapter 7 I discussed our belief in the goodness of economic and population growth. We are hooked on it. As things are, we have to have it. Now our economy must grow just to maintain the status quo. When it does not grow, a number of things happen. Many people lose their jobs—perhaps because we do things more efficiently. Our government's income does not keep up with its expenditures, necessitating cuts in services. Ambitious people feel stifled by the shortage of opportunities. Greedy people are upset by their difficulties in increasing their wealth. And our nation's importance in the world economy diminishes. People cry for action. It is easy to understand why businesspeople and politicians see the economy, and its growth, as more important than anything else except defense.

A company that does not grow may not survive in a dog-eat-dog world. Hospitals, universities, and bureaucracies must expand just to maintain status—if not to increase it. In government and business, a department manager benefits by having more people to manage even when he could get along with less. Most of the time people who inspire growth are noticed, admired, well paid, and promoted.

There are good reasons to restrain growth, however, and perhaps, as we learn more, even to reverse it. A continually expanding economy increases the rate of crowding, resource depletion, pollution of all types, loss of productive soils due to urban sprawl and erosion, and reduction in biological diversity, all adding up to potential worldwide ecological disaster. Continual exponential growth is mathematically impossible in a confined space. In time it must cease.

When a population has expanded beyond the systems that sustain it, further disintegration of living conditions cannot be prevented without outside relief of food and other essentials, and even that will only delay the reckoning. We are actually headed toward such a situation and have no intention of turning back.

To grow, or not to grow? The choice is not a pleasant one. Business, government, and most people push for more growth, making this a nonquestion. Politicians get elected by promising growth, and corporate executives are kept in office by producing it. Who among us knows how economic growth can be ended in an acceptable way, or how we could learn to live with stability?

Oddly the advocates of growth have not asked themselves, nor have they been asked by others, to describe what will happen when growth eventually comes to an end—as it must. The longer we wait before putting an end to growth, the more unpleasant the ensuing situation will be and the worse the problems people will have to live with.

Big!

We all want the biggest, the best, the most—big league ball team, world class symphony, big office buildings downtown, and huge shopping centers—and lots of them. I do not understand how having more and more of pretty much the same shopping centers improves my life. But all this seems to be accepted, without question, as good. Ancient Athens and fifteenth-century Florence were small and had none of these. Detroit and Los Angeles are big and have them all—and more.

The beautiful countryside outside Milwaukee where I used to ride my bicycle is now covered by subdivisions and parking lots. A bicycle ride to the country is now long, dangerous, and unpleasant. Formerly, owning a small cottage on a nearby lake wasn't expensive, but few people can afford to buy one now. When I was a child and my family visited Yellowstone National Park, it was not crowded. The west was still "the west," remote, rugged, with vast lonely places and its own special flavor. It was not like today's duplication of the east, with distant mountains nearly hidden behind superhighways, junk food restaurants, and shopping centers. We were poorer then, without our big time, big league, and more of everything. But we were not less happy. I wonder if other people think about that. I liked that America better than the one we have now. I stand on the roadside looking at what I see around me, and wonder what people are thinking when they say, "I am proud to be American." Are they proud of what we have done to our land?

I do not think our country will be better when there are twice as many people and everyone consumes twice as much as we do now. There will be only half as much space and even fewer resources for each of us. Yet this seems to be what we are working very hard to achieve. Consuming and having more of everything is neither particularly good for us nor for our country. John Herz noted, "There are those who see the most significant developments of our times in the steady and vigorous rise of general standards of living. This rise has produced, and is bound to continue to produce, increasingly an almost exclusive interest of people in their personal well-being, and consequently, a decrease in their ideological and political involvement."[2] The prosperity we have enjoyed in recent years has been accompanied by lower voter turnout, and in my observation, decreased concern for the well-being of the community, its less fortunate members, and environmental matters like energy conservation.

WHAT CONTROLS BIG BUSINESS?

In a privately owned business the owners and managers are one, making it easy to put into practice ethical and other personal beliefs about how to run an enterprise. While it is essential to make a profit, considerations such as pride in the product, employee welfare, and community concern can play an important part in what is done.

Large corporations are not run by their owners. Their owners, the stockholders (which includes pension funds, foundations, and endowments as well as individuals), confine their interest to corporate profits. Actually, many are not even interested in that. Instead, they care only about how much they themselves will make at an acceptable risk over a period of time. As long as their investment makes money for them, they do not care what the corporation does or whether it is disbanded. When they see a better opportunity, they move their money elsewhere. Many enterprises find that the majority of their owners hold their stock for less than a year.

Corporations are run by officers and boards of directors. They collect proxy votes from disinterested owners, reelect themselves to office, and award themselves astounding salaries and lucrative perks. The owners do not care as long as profits are good or share prices rise. Management must make it *appear* that profits will continue to increase over the next year or two. They have little incentive to care what happens beyond that. Even though they may wish to address serious long-term goals, it is short-term results by which investors judge managers.

Executives can be voted out of office or even be sued for not producing acceptable profits. Sometimes at stockholders' meetings, motions are introduced to restrict or eliminate corporate charitable donations (which are largely made for public relations) because they cut into profits. Since large corporations have bought out or replaced many thousands of locally owned businesses that supported local civic projects, this has seriously affected many communities.

This pressure to maximize profits, to endlessly expand, and to compete internationally demands the production of lots of goods or services at low cost. The price for doing this is high and includes lost jobs through downsizing or shifting of the workforce, insecurity and fear among those who retain their jobs; and coercing governments to lower environmental and employee safety protection standards. Even the future of the company may suffer.

Corporations damage communities by moving people in and out and detaching them from their locality and extended families. They remove ownership and management, and sometimes whole businesses, to distant locations. National chain stores cause local businesses to close and replace publicly owned "downtowns" with anonymously owned shopping malls where local citizens' only input is through their pocketbook. Hometown talent and local creations have largely been killed off by "big time" piped in from somewhere else. Costs such as these rarely play a part in decisions made by these corporations. Profits are what count, and the costs are borne by others.

These problems have moved into agriculture. Family farmers have an interest in preserving the quality of the soil for their children. More and more land is now owned by large corporations whose officers are interested in producing profit during their tenure—particularly for the next annual report. They have little incentive to protect the soil for posterity. Much American farmland and aquifers that are rapidly being depleted and are not replenishing are now forced to produce quick profits at the cost of future generations. A large and expanding part of the world's economy is governed by damaging feedback loops like this that produce widespread destruction.

THE INFLUENCE OF BUSINESS

Business has a profound influence on American society. Because it is successful and rich it is respected, admired, and given credit for possessing capacities beyond what it actually has. Its wealthy leaders are listened to by government and citizens alike. To further its influence, it employs experts in public relations, advertising, and lobbying. They are all too good at projecting the limited and often selfish viewpoints of their clients. They promote materialism, consumerism, growth, the virtues of capitalism, minimum government regulation, and low taxes.

Businesspeople listen too, and so recently books, articles, and lectures attacking environmentalists as "prophets of gloom and doom" have become popular. Some of the assertions are correct, but basic premises are generally twisted or far off the mark. Their authors are often associated with conservative groups and make a good living telling businesspeople and industrialists what they want to hear. These pundits may be sincere in their beliefs, but their association with special interest groups and their lucrative earnings raise questions about their motivation. The effect of these often charming people makes it even more difficult to convince businesses and politicians that environmental problems are serious.

Businesspeople and others do not want to hear news that disturbs their ways of doing things, comfortable lifestyle, or ethical complacency. They eagerly believe a single scientist, or someone with no scientific background at all, if she assures them that everything is fine just as it is.

Historically, business has resisted almost all progressive legislation meant to protect employees, the public, or the environment. On the other hand, it has won for itself and managed to retain many advantageous subsidies, and has kept the military-industrial complex alive and healthy. Business has a one-track mind: With great ferocity, it fights anything it sees as a threat.

Oil and coal companies have spent millions of dollars to convince the public that global warming does not exist. They have succeeded in confusing the issue and keeping it off of the public agenda. Particularly misleading are environmental statements from organizations with names like "The Information Council on the Environment," which was created by a group of utility and coal companies.[3]

Industry influences government directly as well. In his book, *The Heat Is On,* journalist Ross Gelbspan describes how in May 1996, Congressman Robert Walker, Chairman of the House Science Committee, disregarded the strong recommendation of the National Research Council to continue full funding for NASA's program to continue monitoring changes in the earth's climate. Instead he turned to the conservative Marshall Institute, "which conducts no original research itself and whose reports are viewed by the vast majority of scientists as political statements rather than as research contributions" for guidance. The Institute denied that there was a crisis and funding was cut.[4]

Advertising Works!

Advertising provides a valuable service. It informs us about the availability and characteristics of products, services, and ideas. However, because it can produce high profits and affect public opinion, it is often misused. Such misuse affects people, society, and, in the end, our planet.

To counter this criticism, advertisers have given themselves respectability by convincing us (they are good at this) that our prosperity depends on their ability to create demands for goods and services. To keep the economy afloat, they must persuade us to buy things that we could get along very well without. This does create jobs, but it also accelerates the depletion of nonrenewable resources and the creation of waste and pollution.

Advertisers tell us that we must produce more in order to keep people busy. A better alternative would be to do this by paying people to do many needed tasks that we are now unwilling to pay for. We would benefit by having more and better qualified people teaching, taking care of the sick and the elderly, and maintaining things, such as roads, buildings, parks, and appliances.

The techniques used by advertisers are almost irresistible because they bypass reason and appeal to our strong primitive urges. Generating wants in people who cannot afford to fulfill them creates a sense of destitution and frustration. Depriving their children of their basic needs or committing crimes may be the only way that some people can satisfy these desires.

Advertising and the distractions it offers captures and fills a sizable portion of our time and memory, not to mention our living space, with junk. This confuses our attempts to judge clearly what is best for ourselves and our country. The ills of society, the fate of our planet, and the welfare of future generations have no place in the la-la land we are continually enticed to enter.

The attractions advertisers offer do not come cheap. Businesses strive to make a certain percentage of profit based on the costs of production, which include advertising costs. When the profits of each member of the chain from producer to retailer are added in, the cost to the customer can grow substantially. For example, in 1994 business and other interests spent $1,547 per 2.7 person household for advertising.[5] If there is a 40 percent profit from producer through to retailer, costs from advertising could go from $1,547 to $2,165. Advertising enjoys our respect but does us considerable harm. We avoid relating it to the waste, pollution, crime, and distorted values it brings about. We defend broadcasters' right to clutter our children's minds with whatever suits their purposes.

Advertising sells political candidates as effectively as it does soft drinks. By making itself useful to businesspeople, politicians, government, causes, and the media, and by providing "free" entertainment, the industry makes many friends. This leaves few with the influence or power to fight against it. The industry is diverse, a headless manipulator with thousands of single-minded purposes. Fighting it is like fighting a swarm of bees with a rifle. However, as we shall later learn, there are things that can be done.

OUR NEW, SYNTHETIC REALITY

For most people today, survival and success depend on getting along in the manufactured world. We need to know how to handle money, understand laws, utilize technology, support our families, and be versed in the social graces. The human-made world is the reality in which most people in the developed nations live. However, species' survival still depends on getting along in the broader reality of nature. While people in the developed nations have many reasons to behave sensibly in the manufactured environment, they have few incentives to get along with nature. What motivates us today does not necessarily further the survival of our species, in fact it often does the opposite.

Indigenous people knew what they needed to know to survive in their surroundings. We are earth illiterates. Few of us have direct contact with nature. We drive through it, fly over it, exploit it, and look at it from comfortable places. We vacation in far-off locales with pleasant climates. We learn much about the shops and restaurants, but little about the natural surroundings or the cultures of the people in these places.

Knowledge of plants, animals, and farming has been supplanted with information about our vocations, politics, products, and celebrities. We know little about how we affect the earth. For example: How many square inches of a tropical rainforest are destroyed to produce one hamburger? How much of a West Virginia mountaintop is ruined in order to provide the electricity to run our air-conditioner? What happens to the paint thinner we pour down the drain? How does the loss of songbirds affect us? We see other people, their political systems, and a deteriorating economy as our greatest dangers, instead of the slow but sure destruction of our life support system.

Our Packaged Lifestyle

The contemporary North American suburban lifestyle ceaselessly consumes land, energy, and resources, turning them into waste and pollution. It demands agricultural and forest products from around the world, causing forests, wetlands, and many species to disappear. This luxurious way of living has become the model for the less affluent in the wealthy nations and for all people in the world's underdeveloped nations.

Consider the way we shop for groceries. Grocery shopping used to be done on foot at nearby stores. Travel to them neither consumed petroleum nor produced air pollution. Apartments or offices were customarily built above stores. Today, the same facilities, each built separately, with all necessary parking, use between ten to fifty times as much land to serve the same purposes. Food stores are built far apart so that people must drive considerable distances over huge amounts of additional land paved as streets in order to reach them.

Pavement begets problems. It supplants oxygen-producing vegetation and in summer adds to heat build-up in urban areas, increasing the demand for air-conditioning. Instead of allowing rainwater to replenish aquifers, pavement quickly discharges polluted water into sewers and streams, increasing flood problems. Parking lots and streets, surfaced with materials requiring considerable energy to manufacture, require electricity to light them.

As petroleum reserves near depletion, or when nations finally decide to limit the combustion of fossil fuels in order to protect the earth's climate, the use of the automobile will probably decrease. It will then be difficult to reach suburban shopping centers, offices, and industrial parks by public transit from low-density residential areas. Because of the lost topsoil, it will be impossible to return these land-hungry complexes and subdivisions back to productive agricultural use.

This type of real estate development is encouraged by local regulations, tax structures, business practices, lending policies, and people's ideas of what is desirable. Retailers are aware of the savings in distribution and operating costs these large centralized locations provide. However, shoppers and municipalities usually fail to consider costs hidden in transportation, time, environmental degradation, and necessary public improvements along with their maintenance. Result: The real cost of a loaf of bread is higher than we think. Part of its cost is paid in taxes, wasted time, squandered resources, and increased air pollution. These factors play no part in current urban planning regarding the location of shopping and services.

It is impossible to live in American cities without causing considerable environmental damage. Most people cannot even get to work or to shops without driving miles in an automobile. Professionals, businesspeople, and many craftspeople simply could not function and compete without an automobile and without providing their employees with air-conditioned spaces. We use our cars to escape the places we have ruined, and in so doing we damage these places further. Suburban adults leave their homes in sealed automobiles, meeting only other cars as they travel. There is no place to walk to, and often no sidewalk to use. For exercise, people drive to health spas where they work out on walking machines. Once people shopped in stores flanking public ways, owned by people they knew. Sometimes they would meet acquaintances on the street. People now shop in chain stores lining malls owned by anonymous corporations. They rarely see anyone they know. Real acquaintances have largely been replaced by the comradeship of television characters.

Everything in our world today is contrived and manipulated in order to part us from our money, and we love it. We have become eager receptacles for thoughts skillfully designed for us by the artists of commerce; we have become good, obedient consumers and are persuaded that our choice of brand is the expression of our personality.

In accepting all of the goodies we have been taught to want, we have forgotten the things that once tied us to nature and furthered the survival of our species in a world still governed by the laws of physics and biology.

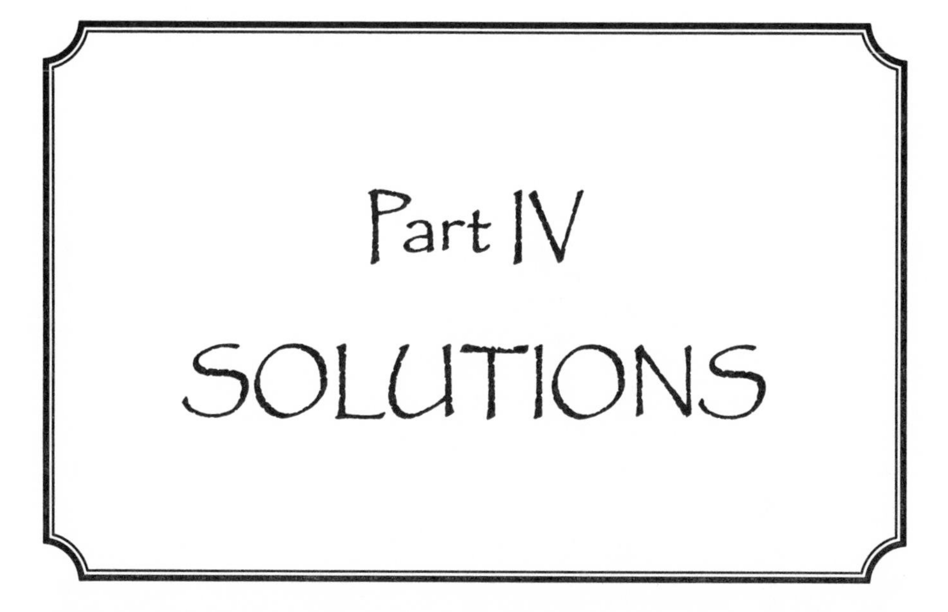

Part IV
SOLUTIONS

An awareness of problems and knowing how to solve them serves little purpose unless we go the rest of the way and actually correct them. Our minds, libraries, files, and data banks are filled with detailed information about problems and suggestions of what we should do about them—and still those problems persist. Besides having a clear idea of what needs to be fixed, we need a strategy, a spark to get things going.

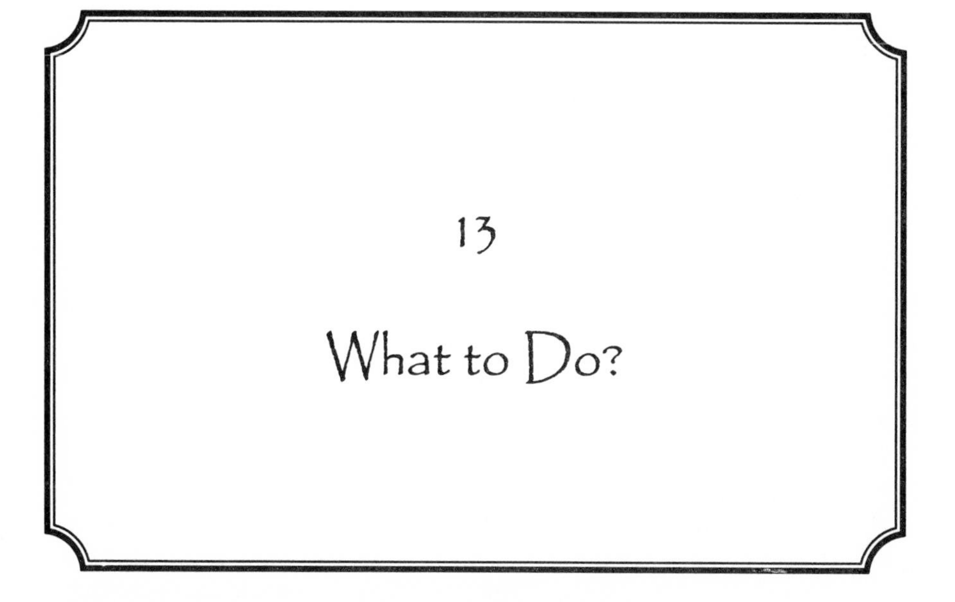

13

What to Do?

Now that we at last can see the invisible walls all around us, we can go about pulling them down. There is no one place to start. We must start everywhere!

As noted in the previous chapters, walls lie within ourselves, within society, and within governments. They involve ethics, goals, and values; how we think; what we think; and what we do. Let's start by looking at ourselves.

KNOW THYSELF

We like to mouth the words attributed to the Delphic Oracle, but we do not take them to heart. If we could manage to know ourselves—to deal with ourselves as we actually are instead of as we wish we were—we could do a better job of getting along with each other and nature. As Jonas Salk put it,

> Perhaps the essence of the crisis of our times is that we are approaching the limit and usefulness of our knowledge of the

> cosmos and are now in need of turning our attention to consciousness of ourselves.
>
> We need mirrors and magnifiers; we need to develop new ways of seeing and of recognizing ourselves. Consciousness of self and consciousness of the new reality are both necessary, and both of the highest value for survival and evolution.[1]

Ervin Laszlo likewise urges us to get to know ourselves better. "It is high time we engage in an individual and collective soul-searching; in a much needed psychoanalysis of our inner limits. Even if the process is painful, its potential benefits should encourage us to carry it through."[2]

Psychologist Robert Ornstein offers this encouragement: "Knowing about how our physiology disposes us toward life may provide us with a way to enact real change that is in line with our character, rather than . . . some artificial or arbitrary system. The brain and the nervous system are not immutable: they grow and change with life experiences. Thus we can take an active role in changing our own brain processes and improve the way we manage ourselves."[3] When we can do this, we can do a better job of confronting and dealing with many problems we face today.

How We Think

No cabinetmaker would build a chest without understanding the capabilities and limitations of his tools. Yet every day, important decisions are made by people who know little about the workings and limitations of their minds. Their ignorance affects the decisions they make and the future of humanity. We need a better understanding of our minds before we can do a better job of thinking and managing our affairs. Knowing more about our deficiencies, we will not be so arrogant about our observations, reasoning power, and memory. Recognizing our limitations would give us a chance to work around them.

We Must Think More Clearly

In the relatively simple environment in which we evolved, it was not necessary to think clearly about complicated things. In fact, sometimes fuzzy thinking and mistaken beliefs helped—as was the case when people made useful discoveries based on faulty assumptions and logic. Consequently, clear thinking is not one of our strengths. Now however, with the powerful impact we have on the earth, poorly made decisions can be disastrous. A better understanding of our brains will show us that our senses, memory, and judgment are not to be trusted. They often play tricks on us. We need to evaluate information better by recognizing our limitations before we put the information to use.

Likewise, an awareness of the power our emotions have over us and of the poor connections between different parts of our brains can help us. It can give us the incentive to make a concerted effort to act more consistently and use our reasoning better to overcome emotions when they start to lead us astray. It is our ability to do things like this, rather than just react to circumstances and situations, that differentiates us from other animals.

Neurologist Michael Gazzaniga is encouraging: "The quintessential human property of mind—rational processes—can occasionally override our more primitive beliefs. It isn't easy, but when it occurs, it represents our finest achievement."[4] One interesting suggestion that could help here is Kenneth Boulding's "theory of bad decisions."[5] Analyzing why we are often drawn to poor choices could help us avoid them.

The huge amounts of data we must deal with today often confuse us. As I pointed out in chapter 3, even in simple societies we rely on defaults, routines, and labels to handle large amounts of information and to deal with the many tasks we face every day. These tools can lead us into trouble, though. People and organizations often take advantage of our dependence on defaults and routines to further their own ends: Bigots engender hatred, hucksters sell us things we don't really need, and politicians create false impressions about themselves or their opponents. We must be taught why and how we use these mental shortcuts so we can recognize and use what is helpful, avoid what is damaging, and be constantly on guard against being misled by inappropriate defaults, irrational routines, and deluding labels.

We also need to remember that we have a penchant to believe despite meager evidence, that we desire simple answers to complex questions, and the consequences of these traits. Acknowledging our own gullibility and questioning information we receive would reduce our vulnerability to despots, charlatans, and emotion-arousing advertising. Knowing the techniques used to manipulate us would better make us aware when this is happening and help us resist. Public awareness of the dangers of fanaticism and how we can be sucked into it would diminish the number of people attracted to this hazard.

As well as striving to override our weaknesses, we should reinforce our strengths. One of our most powerful tools for analyzing and solving problems is intuition. It is usually faster than reason and can draw conclusions on incomplete and confusing data. While it can lead us astray, good intuition can be more reliable than reasoning, which may be faulty. We need to develop our ability to use our intuition effectively, and by monitoring it as best we can with our power of reason, avoid its pitfalls.

Another often undervalued asset is imagination. True, it can draw us away from reality. However, a well-ordered imagination can do just the opposite. It enables us to visualize things we do not experience ourselves, possible futures, and "what if" situations. Very few of us have an imagination capable of constructing a balanced view of reality. When we read that the world's population will soon double, we are largely unaffected because we cannot picture what that means. A good imagination is essential for leaders, although it seems that they often lack it. By being more aware of the value of imagination and further developing our limited supply of it, we can compensate for this shortcoming.

Clear thinking also requires a healthy mind. Children from dysfunctional families often develop distorted views of themselves and reality which can dangerously affect the behavior of whole nations. Society gains by encouraging families to provide a secure, warm environment for children to grow up in. This helps produce adults who react to needs in responsible, rational ways. It reduces the number of people who might grow up with ambitions to be a Stalin or Hitler or who would fall right into step behind someone like them.

We Must Do a Better Job of Learning

We must do a better job of learning and be tolerant and open if we are to gain new knowledge. When we argue, we want to win, not learn. This is particularly harmful in politics, where in the minds of politicians, journalists, and the public, winning seems to be everything and learning, nothing at all. We must be made to see the advantages of seeking truth rather than manipulating preconceptions.

As I have described in earlier chapters, there are reasons we instinctively fear and reject unfamiliar ideas—world federalism, for example. We can change this. We should all be taught to train ourselves to examine new ideas and unconventional opinions for their validity before simply rejecting them out of hand.

How We Handle Data

Information is power. Novelist Doris Lessing noted, "It is information that will set people free from blind loyalties, obedience to slogans, rhetoric, leaders, group emotions."[6] Misinformation and misused information can do the opposite, however, and information ignored will do us no good at all.

What we need today is not so much *more* data, but the ability to locate, retrieve, select, and evaluate what is already available. We need to learn how to access what is pertinent, judge its importance, see how it relates to other things, get it to where it is needed, and use it. This would eliminate much of the waste and apathy of our times. We must find ways to direct data that is critical for human welfare into the hands of the public and decision makers and get them to understand its significance.

One possibility for doing this might be a set of cleverly worked out exercises for manipulating data, which could be seen as games. These could be presented to students and sold in stores. In a challenging, entertaining way, this could give them some practical experience with utilizing information correctly.

We are deluged with trivia today—and often relish it. We need to find out why we choose to fill our minds in this way, giving such data undue weight when there is so much else that desperately needs our attention, such as listening to those who present serious thoughts important to our survival.

We Need to See the Big Picture

By resisting our tendency to view the world as fragments and instead seeing it as an integrated whole, we can deal with problems in their totality, including their sources and broad effects. We can then overcome the causes of various problems instead of responding to symptoms (which often just creates new problems).

People trained to keep an unrelenting eye on the big picture would be invaluable. With our love of specialists, we could designate them "professional generalists." Ervin Laszlo suggests, "We need to complement specialists, who know more and more about less and less, with . . . generalists who know just enough about everything to be able to see the whole forest and not just a multiplicity of trees, and who can therefore be better relied upon to guide our steps and crossroads along our way."[7]

Looking at the whole means asking and dealing with nonquestions, such as, What will happen when the world's petroleum runs out and no inexpensive substitute has been found for the private automobile? When there is not enough food for everyone, how will we deal with shortages? When we need more farmland what will we do about golf courses, paved areas, and low-density suburban subdivisions? What will we do if we cannot provide as much energy as we need, as cheaply as we need to? If we must stop population growth and even reverse it, who is to reproduce and how will this be controlled? How will we deal with the elderly and the large numbers of people with undesirable genetic mutations? How will we deal with terrorism in an increasingly chaotic and vulnerable world? The panel of respected thinkers I will describe in chapter 15 should force important people to publicly address these questions.

Evolution does not look to the future; all species evolved by interacting with existing conditions. Many species cannot adapt fast enough to adjust to today's rapid change and are therefore facing extinction—a fate that could in time be ours as well. To prevent this we must go beyond what our instincts prompt us to do and consciously plan for the long-range future. We can do this. We are the only creature we know of that can think much beyond the next few minutes. We can project, anticipate, and plan. We constantly do this for the next few days, weeks or years; now we must do so for much longer periods of time.

For example, professional societies could set up committees to study the future as it is affected by their professions. They should then find ways that their profession could improve survivability and report this to their organizations. Where they see the need for changes outside their profession, they should note that too and draw the appropriate parties' and the public's attention to it.

We do have "professional futurists" today, but most of them work in limited areas such as the future of religion or international corporations. We need people who study the future as a broad, open-ended subject. Governments should set up agencies to do this and report back to officials and the public.

How We Interact

By acquiring a rudimentary understanding of how memes, which were discussed in chapter 5, spread though a society, and of the effect they have on it, we could reduce the damage that some beliefs are causing today. By being aware of this, people would see how memes that encourage unnecessary consumption, for example, propagate themselves and harm the environment and other human beings. One researcher suggests that being aware of memes will immunize people against injurious ideologies.[8]

Today, we should often act in ways that run counter to our basic drives. A better understanding of how we instinctively react to certain situations would help us overcome unreasonable fears and prejudices that can lead to persecution, atrocities, and war. Learning more about how we interact with others, too, will give us a better understanding of why hostilities break out so easily among us. Knowledge of how we act as parts of groups or crowds would awaken us to our tendency to be overcome by irrationality, hysteria, and inconsistency and show us the consequences such mass behaviors can engender. Classroom games, where players are given a situation and assigned different roles in it, could effectively demonstrate some of these qualities in ourselves.

A better understanding of history would show us what happens when we let our nastier traits govern our relationships with other groups and nations. Genocide and cruelties committed in colonial Mexico, Turkish Armenia, Cambodia, Rwanda, Bosnia, and even in the United States make it clear that barbarity is not the characteristic of any one ethnic group, religion, or period of time, but lies within all of us. By recognizing this, we can make a conscious effort to keep it under control.

History also teaches the high cost of starting a war, as the Germans learned from World War II. We would see how injustice brought about revolutions in France, Russia, and China, and led to wars or conflicts with terrible results in Israel, Iran, Northern Ireland, El Salvador, and many other places. The extravagances of the monarchy brought on the French Revolution, and the mistreatment of Serbs and others helped lead to "ethnic cleansing" in the former Yugoslavia.

People should learn the shameful parts of their own history, not just of others', as is now the case. In the United States, children should know how we treated the Native Americans, for example. When people are aware of their own transgressions as well as those of others, they will feel less righteous in instigating hostilities.

We must learn that our real enemies are not the Serbs, Tutsis, Muslims, or Catholics, but those among us who because of their own ambition, prejudice, hate, or love of weaponry and conflict rouse us in order to further their own distorted goals. We need to recognize that the worst threats we face today do not come from other nations or people, but are the result of the damage we are inflicting on our planet. The consequences of this will be far greater and more final than anything done to us by other people. An awareness of this should pull us together to fight this common enemy.

We Need to Develop Means for Bridging Our Limits

We cannot exchange our brains, but we can improve the way we use them. We can develop and use techniques and tools that supplement our brains, override their quirks, and enable us to do things we normally cannot. We already do this in many ways. We use a telephone to talk farther than we can shout, writing to keep records, and police to maintain order on the streets. We use tools and organizations to do things individuals and groups cannot. Accounting, environmental impact studies, data banks, and orbiting satellites, when used effectively, can greatly extend our capabilities. The scientific method, management systems, the systems approach, operations research, brainstorming, formalized techniques for constructively resolving conflicts, and computers discipline and expand the capacities of our minds. An interesting example of this is given by Irving Janus, who offers suggestions for countering "groupthink." He suggests that "a group designate one member as a deviant—that is, as a critical evaluator of what goes on, raising objections and doubts. The Devil's advocate can save the group from itself, making sure it faces uncomfortable facts and considers unpopular views, any of which could be critical for a sound decision."[9]

We have the potential to do far more along these lines. We can develop new tools and techniques that can eclipse our natural inclinations and guide us in constructive, safe directions.

WHAT WE THINK

Surgeons not only need to be able to use their instruments skillfully, they need to understand the human body. Likewise, besides using our brains effectively, we need to understand the world and the problems we face. While we cannot expect everyone to have an in-depth knowledge of psychology, sociology, politics, physics, and biology, for example, there are specific things in these areas everyone should know. I will discuss this in detail in chapter 15.

We Need a Good World Model

Each of us needs to inspect and revise our inner models of the outside world. They need to be brought closer to the way things are and away from mistaken ideas, biases, unfounded fears, and wishful thinking. Poorly grounded models can produce beliefs that work against our survival, for example, that we are superior, birth control is a sin, growth is good, evolution is wrong, and technology will solve everything. Although we must protect the right of individuals to hold differing beliefs, where these beliefs harm others and threaten our future they must be confronted. Their fallacies and the threats they pose must be exposed. Simply dealing with the harm that results from the beliefs—treating only the symptoms—is not enough. Beliefs need to be examined, compared with established facts, and our world models adjusted accordingly. An awareness of our mental limitations, too, as discussed earlier, can help us see the voids and inaccuracies in our world models so that we can compensate for them.

Every age, nation, and peer group has its blind spots, unexamined beliefs, and nonquestions that are out of line with reality or the long-range interests of our species. For instance, it was once accepted that the earth was the center of the universe. It has not been a custom of peoples to look for their blind spots. We must make it a habit to look for them and compensate for them. Appropriate nonprofit organizations could decide to contribute to society by seeking out and investigating blind spots, unexamined beliefs, and nonquestions and then bring them to our attention.

Improving our models requires a better understanding not only of ourselves, but of the world. We must recognize that we are part of nature and dependent on natural laws and systems that maintain the narrow range of temperature and pressure we need to protect us from dangerous chemicals and radiation. We need to recognize that all things are in some way interconnected and that one cannot do something in one place without affecting other things.

A world model which mimics reality as closely as possible, produced by a panel of respected international scholars and accepted by most nations, would establish a common ground among people. It would reduce conflicts and enable us to focus our limited energy on meaningful problems. Constructing a very simple world model as a classroom project in junior high school would give students a basic understanding of important relationships that would remain with them throughout their lives.

Instead of seeing a piece of land as something to own or exploit, we should view it as many Native Americans did—as a living organism, the source of all we need—and recognize that we are every bit as dependent on the soil for our existence as a rose bush is. We need to understand that we are not complete and self-contained but depend on other living things and the minerals and atmosphere of our planet as well as energy supplied from a distant star. Viewing ourselves this way, we would see that our lives are not unique but rather a link in a chain originating in the past and hopefully continuing into the future. We would want to learn about the wonders of our universe and about all forms of life, and we would respect all living things.

We need to gain a better understanding of something the majority of the world's people and their leaders have little interest in: the limits of the earth's ability to continue supporting the demands we place on it. This would encourage us to learn how to achieve our ends more efficiently with less waste—and be more moderate in the demands we make in the first place. An interesting project for elementary and high-school students would be to construct lists of what could be done to reduce our burden on the earth.

One of the most important laws that govern life on earth is that of evolution. Living without understanding this today is like practicing law without being able to read, except that the consequences can be far worse. Father Thomas Berry sees a religious need for understanding evolution. "The time has come for the most significant change that Christian spirituality has yet experienced, but this change is itself part of a much more comprehensive change in human consciousness brought about by the discovery of the evolutionary process."[10] By accepting and understanding evolution, Christianity, which Berry sees as a process, moves forward—which it must.

We will also benefit by recognizing that the laws of evolution apply to human society and culture as well as to blackberries and bumblebees. Groups that rely on fallacious information, for example, are less likely to survive than groups that utilize the truth. "Understanding these laws might help us to resolve current socio-cultural crises and positively influence the future."[11]

Everyone should have a rudimentary understanding of ecology as well. A man sitting on a branch realizes the branch is important to him, so he will not saw it off. People who understand their dependence on nature are less likely to destroy it.

ETHICS

Life would be simpler and it would be much easier to resolve our problems if everyone adhered to the same ethical principles. As we saw in chapter 8, this is not the case. We live in a shrinking world full of conflicting ideas and convictions, clashes of value and belief. Demagogues and opportunists profit by stirring up the hostilities that can easily result. The rest of us gain by overcoming these differences. We must do so if we want to survive.

Universal Ethic Needed—And Possible

Ervin Laszlo observes, "Truly universal values do exist. They lie at the core of all major religions and our most noble cultural traditions. The values of universal brotherhood, love for one's neighbor, and the golden rule of treating others as we ourselves would wish to be treated are just some of the ideals that are common to all cultures."[12] Philosopher Paul Kurtz expands on this, suggesting, "We should seek to transform a blind and conscious morality into a rationally based one, retaining the best wisdom of the past but devising new ethical principles and judging them by their consequences and testing them in the context of lived experience."[13]

Respected religious leaders and ethical scholars from different countries and cultures need to work together to produce a set of ethical principles that most people could endorse. While religious leaders differ on what God is and the purpose of life, they do have common interests that should induce them to cooperate. Widespread publicity given to these agreed-upon principles would affect public opinion, put pressure on political leaders, and head off much of the conflict between the people of the world.

Ethics is not only a theological concern, as some religious people might think. It is philosophical and practical as well. Democracy, business, science, and society cannot function without it. There are many people who were not raised in any religion or who do not follow one in their adult lives. They do not have a formal ethic to follow, providing additional reasons for the creation of a universal ethic that all people can accept.

What Would a Universal Ethic Include?

Such a set of ethics should be very basic and involve only those things that could be widely agreed upon; for example prohibiting unjustified killing, harming another person or nature without justification, stealing, and lying. These principles would apply not only to our one-on-one dealings with other people, but the relationships of governments and those they govern, our relationships with other nations and their people, and international business activities. Our "universal code" would include consideration for future generations, too, and guidelines for prudent use of our planet's natural resources.

Do Not Neglect Nature

The money and heirlooms we leave our descendants will do them little good on an overpopulated, impoverished planet. We must come to realize that if we care about our families, we must also care about nature. Treating nature ethically becomes one with our concern for those who come after us.

Animistic religions hold the earth and all forms of life to be sacred. The Native American people, for example, have "one of the most integral forms of spirituality known to us. The cosmic, human, and divine are present to one another in a way that is unique. . . . This is precisely the mystique that is of utmost necessity at the present time."[14] These Western thinkers have expressed similar thoughts:

> In relations of humans with the animals, with the flowers, with the objects of creation, there is a whole great ethic scarcely seen as yet, but which will eventually break through into the light and be the corollary and the complement to human ethics. . . . Doubtless it was first of all necessary to civilize man in relation to his fellow men. With this one must begin and the various lawmakers of the human spirit have been right to neglect every other care for this one. That

> task is already much advanced and makes progress daily. But it is also necessary to civilize humans in relationship to nature. There, everything remains to be done.
>
> —Victor Hugo[15]

> What is needed . . . is a new ethic, ideology, or theology that will make it sacrilegious to deplete natural resources, to pollute the environment, to overpopulate, to erase or degrade other species, or to otherwise destroy, demean, or defile the evolving quality of the biosphere.
>
> —Roger Sperry[16]

Evolution has produced an intricate physical environment that includes atmospheric oxygen, the food chain that supports us, hydrocarbons we can use for energy and making things, and ourselves. It is only right that we respect and protect this process and the earth's precious ecological system as parts of the grand design (whoever or whatever is behind it), instead of inflicting damage on them as we are now doing.

Contemporary Western society considers the rights of the individual to take precedence over those of society and the well-being of society above that of nature. This runs counter to reality. If nature is unhealthy, society will decline, and if society does poorly, individuals will suffer. Roger Sperry writes,

> For the first time in human history, global conditions have reached a stage that demands value perspectives which transcend not only innate biological drives but even humanitarian guidelines that have been respected for centuries. What may appear today to be most humane, compassionate and civically and morally upright, may later prove to be most inhumane, cruel, and sinful when viewed from the standpoint of those many hundreds of generations hopefully to come.[17]

It may actually be kinder, for example, to allot limited capital to birth control than to feeding the hungry, because over time it will prevent a larger number of people from starving.

A Kenyan proverb states, "We should think of our resources not as having been left to us by our parents, but as having been loaned to us by our children." Economist Kenneth Boulding said this in another way: "Posterity enables us to repay the debt that we cannot repay to our ancestors. We are inheritors of countries, cities, libraries, universities, a great body of knowledge, buildings, literature and art, all of which we did nothing to produce. . . . So the gifts we have received from the past create a sense of obligation which can only be relieved by gifts to the future."[18] How can one argue against the rightness of these statements? Yet we violate them every second.

Likewise, in our world today it is clearly immoral for some people to live extravagantly while others live in misery and starvation. The affluent sometimes claim that the solution to this lies in the poor raising their standard of living, rather than in a more equal distribution of wealth. However, our planet, with its swelling population, simply cannot support everyone living like middle-class Westerners. Ethics requires that we find a just answer to this pressing nonquestion.

In primitive societies, there was a place and task for everyone. Not so today. There are people who find themselves unneeded and unwanted. Because they are a minority their problems are unimportant to most others. Those of us more talented and energetic, healthier, and luckier should be grateful and use these gifts to make a place in society for those less well off, rather than blaming them for their insufficiencies. Besides this being the right thing to do, there would be an added payoff: People who believe they have a chance to participate in society's rewards are less likely to lie, steal, and kill.

SOCIETY MUST ENCOURAGE ETHICAL BEHAVIOR

Today there are many influences that encourage unethical behavior or trivialize our thoughts. Bigots kindle racial, religious, and ethnic tensions. Advertising extols consumption. The media glorify physical beauty, having fun, acting tough, and impressing others. Some politicians tell us that the pursuit of our own welfare is the best way to help others. The people pushing these views so zealously are richly rewarded for doing so. They pursue their task with dedication and manipulate our emotions to help them. Those who counter this, who must appeal to our powers of reason, are usually fewer in number and do not receive the rewards their opponents do. In the United States, our commitment to separation of church and state makes us reluctant to discuss ethics publicly or teach it in schools. This void is a poor defense against the other voices our children are exposed to.

We must motivate parents, teachers, and the media to cultivate ethical values. Among other things, we need to be taught that ethics are not passive; they require us to act and speak out. Respectable citizens, by failing to do so, may allow terrible things to take place. The maltreatment of child labor in the early Industrial Revolution, the takeover of Germany by the Nazis, and the accelerating eradication of many species are examples. Bystanders have obligations; ignoring them is what Christians call a sin of omission. By ignoring needs and allowing harmful activities to take place, we participate in unethical acts by our indifference or cowardice.

Political scientist John Herz points out that "everybody working in the field of science and technology must realize that he cannot escape moral responsibility for what his inventions, discoveries, and work are used for. Nor can social scientists be satisfied any longer with ivory-tower theorizing and abstractionism."[19] And, "In our age of global survival concerns it should be the primary responsibility of scholars to engage themselves in survival issues. We can no longer afford the luxury of 'value free' research."[20] Organizations of scholars and scientists should set up ethics committees that identify projects considered morally questionable, and notify membership of their concerns. This would place members engaged in such activities in uncomfortable positions.

Oscar Schindler, who has become famous due to Thomas Keneally's book *Schindler's List* and Steven Spielberg's movie of the same name, was a greedy opportunist out to make money. He became caught up in the horrors of the Nazis' "final solution." This became too much for him and a tremendous sense of decency and genuine selfless heroism emerged from within him. Other people who appear to be insensitive to the damage their crassness has brought about must also have this selfless side. Business executives consumed with profits and politicians determined to win at all costs need to recognize that they are human beings with responsibilities extending far beyond their short-term goals. Just as there are dark sides to us, there are also good sides. Finding ways to encourage the good and discourage the bad could help humanity in many ways.

As I pointed out in chapter 3, our behavior is affected by the groups to which we belong. We need to study these groups and pressure their members to encourage constructive rather than destructive behavior. Peers and society in general influence us as well. When society admires power, money, and fame, this is what we want and strive for. Business and the media promote such values, overriding the teachings of moral leaders, teachers, and parents. With the help of people who understand advertising and psychology, we must somehow change these goals.

WE MUST EXPECT AND DEMAND ETHICAL BEHAVIOR

Today it is considered wise to invest your money wherever you can get the highest returns—as long as it is legal. People are admired for doing so. We tolerate and even reward nations, especially our own, that disregard the welfare of others to gain advantage, as long as it does not get too far out of hand. Many Americans approved the United States' funding of covert operations in Cuba because Castro was a communist. A public less tolerant of such things would mean less misbehavior.

We must expect and demand ethical behavior from individuals, organizations, businesses, and governments. In the 1950s, many people told me they carried a five-dollar bill on top of their driver's license to mollify police in case they were stopped. They took this as the normal way of doing things—and in that town, at that time, this was the pattern in numerous facets of life involving local government. In many countries things are even worse. I do not know anyone now personally who talks about having bribed police. My own experiences with government officials have been straightforward. When people expect it to be this way, they have a good chance of getting what they want. In Mexico City, for example, people expect and support corruption, and that is what they get.

The world community, through the United Nations, should go beyond making empty threats and consistently exact a price from national leaders for offenses such as abusing their own people or selling weapons of mass destruction to outlaw nations. Offending leaders and their ministers, for example, should not be allowed entry to other nations and sometimes embargoes should be enacted and enforced. I believe that there is an unwritten conspiracy between most world leaders to be easy on each other (like the aristocrats of yore, who were entirely civil to each other while ordering their soldiers to kill and torture enemy soldiers). This type of behavior is perhaps realpolitik, "being realistic," but seen with clear eyes, it is simply criminal behavior on a grand scale. We should deal with these people accordingly, as best we can, instead of pampering them.

Manufacturing radar detectors or running an advertising agency that promotes waste and superficial values gives one far more prestige than the very useful job of collecting garbage. If society attached a stigma to occupations and activities that hurt us and raised the status of those that benefit us, more of us would turn to socially useful occupations, and politicians and businesspeople would be motivated to behave more constructively. Shedding hypocrisy, in addition to some lucid thinking by all of us, would help here. For example: Shooting someone for his money is clearly considered murder. Is it not also murder to cause death by another means for money? What if that way is producing or selling a product or harming the environment, causing death and calling our reward "profit"? What is the real difference? Knowingly causing death—by gun or the stroke of a pen—should be punished consistently. We should all recognize that by investing in businesses that gain by causing harm, we are participating in this activity. Using the same reasoning, isn't killing for your country when you know it is wrong also murder? By acknowledging these realities, people would behave better and society would have less clouded ethical standards.

We need a public milieu that creates a sense of shame in people who harm society or other people. The prestige one now gains from greed and conspicuous consumption should be turned into an embarrassment. It should become commonly recognized that people who pursue these goals are acting on their baser instincts and injuring us all. Journalists, teachers, entertainers, writers, and artists would have to be involved in creating and promoting this new set of values. Some will say that this is manipulative, but aren't the methods used by those who pull out all the stops to get us to buy manipulative? Don't we manipulate ourselves when we form positive habits? Why allow destructive elements to use weapons we deny ourselves?

We must persist in influencing people, organizations, and nations in every way we can to behave ethically. Our commitment to exerting pressure here must be more than a continuing endeavor, it must become habit and even be institutionalized.

WE MUST PROTECT NOBLE IDEAS

Through human history noble and good ideas have been twisted by opportunists who saw a benefit for themselves in these distortions. The Crusades and the Spanish Inquisition, for example, were clearly events that ran counter to the positive teachings attributed to Christ. In trying to effect a more reasonable and ethical world, we must recognize and face this recurring problem and find a way to deal with it in a general way, rather than coping with each occurrence as it arises. (Until now we have done a poor job of even doing that.) We need to find ways to protect well-intended ideas so that we do not have to deal with the sometimes terrible consequences of their perversion. It's a difficult project, perhaps impossible, but worth trying.

WE NEED REAL GOALS

In other times and cultures, there were wise men and women we could consult about what we should do next and where we should be headed. Today we have experts, people knowledgeable in one area and ignorant in most others. Clerics, economists, political analysts, financial advisors, marriage counselors, career consultants, and psychiatrists are all experts. We are left with almost no one to advise us about overall objectives, how we should direct our lives and efforts. Hucksters have filled this void. In serving their own ends, they leave us floundering in a sea of confusing, disconnected, superficial ideas, beliefs, and goals reinforced by our own primitive urges to seek personal pleasure. Our failure to achieve this (through possessions, power, or fame) leaves us discontent and more likely to participate in antisocial activities.

Currently, the only societal goal that seems to be widely accepted is growth—and ever more of it. Under public pressure, and to avoid falling behind economically, nations promote the pursuit of more and bigger of almost everything. Except for this, industrial societies have no broadly accepted set of values as traditional societies have or had. This makes it hard for us to deal with many problems such as birth control, poor education, or caring for the unfortunate.

We will remain in this quandary until the nations of the world and most of their people formally agree upon some basic goals. Like basic ethical principles, these goals should be limited, but they must address the well-being of our species and our planet.

Throughout evolution, survival has been the overwhelming drive of all species. It must again become ours. The future survival of our species must become more important to us than comfort, possessions, economic growth, and nationalism.

There are two places to start. We must see the absolute need of reinstating survival as our primary goal, and clearly see that consumption, wealth, and beating the other fellow neither serve this goal nor bring real happiness. We should regularly be reminded that people obsessed with material goods are threatening us all and that seekers of money or power are often very limited people with primitive motives and warped outlooks on life. Then goals and status symbols would change. The public would see those promoting unnecessary consumption as unpatriotic and people with goals more in tune with universally accepted ethical principles would be respected.

If we see our real battle as the need to control ourselves in order to harmonize with nature and each other, people might be willing to sacrifice and share a common goal as they did in World Wars I and II. A sense of winning and losing engendered by the media, as is done in sports or a war, would raise public interest and enthusiasm for our battle to protect our future. We must transfer our desire to win from winning over others to winning the battle for survival.

WE MUST CONTROL WHAT WE HAVE CREATED

We all need to better understand the effects of the tools and techniques we have at our disposal on the future of our planet and our species. What are the long-term effects of chemical fertilizers? What are the possible consequences of chemical or germ warfare? How will future generations cope with stored nuclear wastes and their vulnerability to the increasing number of terrorists and possible future chaos? How will urban expansion survive without the huge expenditure of energy we now devote to transportation? We need to analyze the long-term effects of many current practices in isolation as well as in conjunction with each other. This would be a good topic for a classroom exercise.

In recent centuries we have welcomed every marketable invention with open arms, trusting that these things would do wonders for us. Then, as with the automobile, we let them take over and dominate our lives. This does not have to happen. The Japanese samurai restricted the manufacture of guns. Prior to World War II the last battle where firearms were widely used in Japan was in 1637, and by the early eighteenth century guns had become rare and were not a menace. We could do the same with many of our inventions-turned-problems if people would recognize this possibility, agree, and demand that something be done. Eliminating nuclear and other weapons of mass destruction, for example, would make the world a less dangerous place.

WE MUST CONTROL OUR NUMBERS

As we have to gain control of our inventions, we must also curb our numbers. Already our descendants will have less land and less food per person than we do, and a point has or will soon be reached where the biosphere cannot sustain itself into the future under the burden we place on it.

Today, responsible people call for containing the world's population. This means dealing with the desire of people to have as many children as they wish on a finite planet. China, which sees the seriousness of the problem, exercises mandatory controls—and is strongly criticized. Those who object do not offer an alternative to stabilizing our numbers other than volunteerism, which in most undeveloped nations shows little promise of working. Where personal restraint is not working, we will have to create incentives. There are many possibilities for both voluntary and mandated approaches.

Numbers are not all that will be important for future generations. As I discussed in chapter 7, the quality of the genes and parenting we leave them will affect our progeny greatly. This raises difficult questions. What is quality? Who is to judge? How are controls to be implemented? These questions are difficult to answer, but if human posterity is to be tolerable in the future, they must be answered in ways that produce positive results.

If people are not mentally healthy, we cannot expect them to raise a healthy family or meet the needs of society, even when it is in their own interest. Child abuse and poor upbringing repeat themselves from generation to generation. Whereas in primitive societies dysfunctional families had a poorer rate of survival, today many irresponsible, abusive parents produce more children than other people, perpetuating and probably increasing this problem for the future.

We must find democratic, humane ways to discourage people who do not want to assume the responsibilities of loving parenthood from having children. Many people refrain from starting a family until they can support it, yet their taxes go toward supporting the children of people who feel no such responsibility. We do not want children to go unfed and uneducated, but if we are to care for the offspring of these people, we can ask them to practice some form of birth control and attend classes in parenting. If these steps do not work, for the sake of future generations we have to look to sterner measures.

REPLACE COMPETITION WITH COOPERATION

As I mentioned in chapter 4, we spend a huge amount of our time and talent working against each other. This happens all the time, as we all (as individuals or groups) focus on different activities and goals, and over time, we change direction and take a new course, ignoring and undoing much of what we were originally trying to do or had accomplished. By working together we have a gold mine, a tremendous potential for freeing up energy and doing a better job of solving our problems.

If we were to devote just half of our competitive effort to cooperating, we would all gain immensely. Our present confrontational mentality, for example, favored by growing numbers of lawyers, can be so distracting that problem solving often produces distorted results and the common good suffers.

If we are to have a viable future, individuals and society will have to act consistently to coordinate their plans and activities with each other to achieve the maximum overall good. We must find ways for people to work together to solve problems and make a better future without having decision-making and planning degenerate into bickering and politicizing, as now so often happens.

We are all parts, not wholes. If men and women, doers and thinkers, scientists and businesspeople would respect each other and work together constructively, we could accomplish far more than we do now. A football team has a much better chance of making a touchdown when each player isn't trying to do it on his own.

By rewarding teamwork when it replaces harmful competition, we will nurture collaboration and discourage violence. Working together harmoniously does not make headlines or create high TV ratings, but if we valued it more, journalists would give it better coverage and politicians would take note.

Cooperation does have payoffs, and if the compulsive competitors among us could be made to recognize it, it would benefit us all. Richard Dawkins writes,

> Many situations in real life are, as a matter of fact, equivalent to nonzero sum games [games where there are no losers]. Nature often plays the role of 'banker,' and individuals can therefore benefit from one another's success. They do not have to down rivals in order to benefit themselves. Without departing from the fundamental laws of the selfish gene, we can see how cooperation and mutual assistance can flourish even in a basically selfish world.[21]

USE THE BEST MEANS AT OUR DISPOSAL

We have means at our disposal that could help us solve our problems—if we would use them to their best advantage. Instead of doing so, however, we too often expect yet-to-be-developed information or technology to come along and save us. We would do better if we learned to play the instruments we have, rather than wait to add uninvented ones to our orchestra.

We also have access to minds that could help us. However, because they have not devoted themselves to establishing relationships with the powerful, may have unfamiliar ideas, or do not conform to conventional thinking, we do not use them. Novelist Doris Lessing, in discussing people with courage and integrity who avoid being part of the crowd, suggests that "our future, the future of everybody, depends on this minority. . . . We should be thinking of ways to educate our children to strengthen this minority and not, as we mostly do now, to revere the pack."[22]

Doers and leaders who can only think in conventional ways need to find a means of selecting and utilizing people with capabilities quite different from their own. We cannot afford to overlook the good people and ideas we now bypass. People who do well on tests—or have charm and connections—are not necessarily the people who have the foresight, creativity, and imagination needed for many tasks. We have to analyze current ways of choosing people and ideas, and improve them or find better methods of selection. One possible way, not without its own problems, would be to have professional organizations appoint outstanding people in their fields to make up selection committees. These committees would then choose people who would serve as advisors to political leaders, legislative bodies, and government agencies. We also need to develop methods for choosing better elected officials—ones dedicated to solving problems, rather than those whose knowledge and skills are directed toward winning elections and are motivated by primitive drives and a willingness to do most anything to stay in office once they get there.

Above all, we must keep in mind that the most efficient and lasting way to correct the problems we face—environmental, social, or economic—is to eliminate their causes. Society and individuals have benefited enormously from the elimination or control of viruses and bacteria such as smallpox rather than treating each case as it arises. Likewise, we would gain by eliminating or reducing the causes of war and crime rather than dealing with the events themselves. To do this, we must utilize the best minds and ideas we can find.

Urban planning presents one obvious example: To reduce problems caused by the automobile, the basic question should not be how to lure people out of the car and into faster trains, or how to save fuel and reduce pollution by producing more efficient cars. The question should be how to lessen the need or desire to travel—and the distances one must cover to satisfy his needs. Addressing why we go from one place to another will direct us to answers. Some will be in areas such as redesigning cities, others will involve ourselves. What has value, what really makes us happy or adds to our lives?

Knowing that junk food is fattening and deciding to stop eating it does us no good unless we actually cut out the potato chips. How we think and what we think form the foundation for achieving our goals, but what ultimately counts is what we do. We must find workable ways to further constructive behavior and hinder what is harmful, and then put them into practice, which is what we will discuss in the next two chapters.

14

Governments Must Help

Human beings are social animals. Besides enjoying one another's company, we need each other. Without the help of others we can subsist on only a primitive level. Like the honeybee, but less strictly so, we need social organization: a government and a social order to make rules and see that they are followed. Governments can do things that are hard or impossible for individuals to do. They can build streets and libraries, maintain order among people, and protect them from criminals and enemies. They can also remove some of the invisible walls that people acting alone cannot deal with. For governments to do this most effectively, they need to start by looking at their mission from the ground up.

START WITH THE IDEAL

Our political-economic world has been built piece by piece out of dreams, ideas, primitive urges, fears, prejudices, and narrow interests. It is largely a collection of pieces and patches, certainly not the best we could do. When working on town planning projects in my own field, architecture, I find an ideal design to be a good guide. It keeps me aware of what is most desirable, and when compromise is necessary, keeps me headed in the right direction. Working in this way rather than just fixing what we see as wrong, adding where needed, and arguing about the details, would help us get closer to what a better world should be like.

Political scientist Robert C. North describes the benefits of making societal models: "Utopias or abstract models can be used as yardsticks, so to speak, for measuring societies in the real world or as takeoff points for the generation of new alternatives. Utopia building can be useful in helping us to identify our fundamental assumptions, to make operational values more explicit, to identify ways in which basic values conflict, short-term interests damage long-term interests, and so forth."[1] Utopian models could be developed by government planners, consultants, academia, nonprofit organizations, citizen groups, or even individuals.

It would be civil suicide to throw out what we have and replace it with a utopian model. A revolution would be needed to do this. The history of revolutions shows that they cause much human misery and often end in disaster, and in the process much that is good is lost. However the thought that goes into making ideal models, unencumbered by conventional thinking, can give us a broad, clear view of our situation and reveal ideas that can help us clarify where we want to go and help us get there. Let us start with a broad stroke, and then fill in the details.

THE TIME HAS COME FOR GLOBAL ORDER

Ideally, centers of government should be as close to their citizens as possible. This enables officials to best accommodate local conditions and allows people to participate in decision-making most effectively. However, today we face challenges our ancestors did not. We routinely conduct business around the world and have the capacity to annihilate most life on our planet. No country can fulfill the traditional function of nation-states: that of protecting their citizens. Governments can neither shield them from nuclear attack nor spare them environmental damage such as destruction of the ozone layer or depletion of the earth's fish population.

If all nations were like-minded, moral, and had identical interests, their relationships would be mutually beneficial. Since this is not the case, some form of international organization is needed to establish and maintain order. However, nations are unwilling to forego their prerogative to do exactly as they wish. We must come to see that a world divided by such thinking, or by narrow national self-interest, is a world in chaos, just as dangerous as any individual nation would be without a government or justice system.

All the world's people and every sovereign political body should realize that we gain by living harmoniously with each other. This call for an international, worldwide law has been sounded throughout the late twentieth century. Consider the following quotes:

> The world no longer has a choice between force and law. If civilization is to survive, it must choose the rule of law.
>
> —Dwight D. Eisenhower[2]

> We must create world-wide law and law enforcement as we outlaw world-wide war and weapons.
>
> —John F. Kennedy[3]

> We have the opportunity to forge . . . a New World Order, a world where the rule of law, not the law of the jungle, governs the conduct of nations.
>
> —George Bush[4]

> There is no salvation for civilization, or even the human race, other than the creation of a world government.
>
> —Albert Einstein[5]

For worldwide law to develop, nations must be willing to surrender some of their sovereignty and we must do our best to think more rationally about governance, replacing chauvinism with respect for people of all nations.

Order and law must not stop at national boundaries but be extended to include all humanity. This could be accomplished through a loose world federation, operating under enforceable laws, which would not deprive people of local control over most matters. It could further individual freedom by defending people from ruthless tyrants who would like to gain control over them. Such an international confederation would resolve disputes by relying on law and conflict resolution techniques, eliminating war as a means for doing so. It would control environmental matters that extend beyond national boundaries, such as protecting aquatic life in the oceans. It would eliminate pressures on nations to weaken their environmental regulations in order to compete more competitively in global trade. Global security would free large amounts of resources now spent on defense for needs such as improving the lives of the less fortunate.

Another problem that should be dealt with on a global level is unequal access to the earth's resources. Some countries with large, exploding populations have little productive land, forests, or mineral reserves. As I mentioned earlier, most of their people live in misery while residents of other countries bask in luxury, consuming and polluting at a level our planet cannot support.

We avoid thinking about this uncomfortable predicament, but we must address it. Like Native Americans who held mineral resources, soils, and forests to be sacred, we should assume that riches like these do not belong to anyone but should be prudently and justly shared among all people, including those of future generations. A different attitude toward world resources, a generally accepted basic code of ethics, a worldwide organization with enforceable regulations, and the commitment of wealthy nations to help those driven to overexploit their resources would help accomplish this. Reducing the difference between the wealthy and poor would hurt no one and would eliminate some good reasons for conflict.

There should be a worldwide bill of rights that guarantees everyone freedom of religion and speech and access to accurate information. Students must hear truth in their classrooms. They must learn real science and true history, not national, religious, or racial distortions. They have a right to be educated, not indoctrinated. Citizens with an accurate understanding of reality will treat each other and the planet better.

Individuals as well as governments could participate in the establishment of the new global order. John Herz suggests an international civil service as a potential "universal class," above parochial interests.[6] This would be attractive to many young people, provide them with valuable, lifelong information about our world, and establish bonds between peoples, as well as serve other useful purposes.

PUT NATIONALISM IN ITS PLACE

While order is needed on a world level and some things can only be managed there, most government functions deal with the everyday affairs of their citizens. This is why, as noted earlier, government should be as close to the people as possible. A greater awareness of both the benefits of regionalism and of the need for global order by the public, business, and the media would strengthen these levels of governing and weaken the dominance of power now at the national level. This would further peace by reducing dangerous international conflicts and would strengthen democracy by bringing government closer to the people.

Ideally, governments' area of jurisdiction should be established by considering not just the cultures and languages of the people involved and their ability to participate in the political process, but geography and ecological communities. Special interest groups, ultranationalists, and the personalities of political leaders, however, make it exceedingly difficult to change national boundaries, most of which are more the result of historical power grabs than of any good sense. Even so, we can still accomplish much working with nations as they are today by seeing that they act in a more rational, ethical way.

Introduce Reason to Public Affairs

Political leaders do far better when appealing to emotion and basic drives than to reason. Reducing expenditures on rehabilitating convicts, for instance, saves money in the short term, but is expensive later when these people return to society with their bad habits unchanged. But this is and example of what we are likely to hear in an election year.

The operations of our mind, society, government, and the election process interact to create a situation where short-term selfish interests and mind manipulators control the agenda of government. Knowing this, we should be able to redesign some of these mechanisms and structures so that the electoral process better addresses real needs. Perhaps a special (and adequately funded) semi-independent branch of government could present both the pros and cons of various issues to the public in a clear, straightforward, understandable way.

Government-funded political campaigns with strictly enforced procedures established by broadly representative panels of citizens would help. Candidates would be required to address specific topics, as newspapers sometimes now ask them to do. If reason is to prevail in public affairs, clear thinking, responsibility, and truth must be made to bear more weight than emotion, fads, and well-funded interest groups.

THE BIG PICTURE

General Omar Bradley once suggested that we steer "by the stars, not by the lights of each passing ship."[7] Governments too must view and deal with things holistically instead of as unconnected pieces, one at a time. By coordinating government activities we can make them work to enhance each other. Taxes on petroleum, for example, raise money, and by encouraging conservation, make us less dependent on unreliable foreign governments; improve our balance of trade; and reduce automobile accidents, urban noise, and pollution. Such taxes would also reduce sprawl, conserving valuable agricultural land.

When we look at problems, possible future economic collapse and political chaos for example, we tend to isolate them from others such as the costly long-term storage of nuclear waste. By segregating things this way we deal with an unreal world. Governments must, as best they can, study the interactions of all projections and likely possibilities, and incorporate their findings into their plans for the country. They must examine the combined interactions of population growth, climate change, changes in health care, bacteria resistance to antibiotics, new technologies, Gross Domestic Products (GDPs), terrorism, deteriorating human genes, changes in communications, and the depletion of critical natural resources.

When things are changing fast and governments' decisions have an ever greater impact on the future well-being of our species and planet, finding good answers to questions and solutions for problems becomes crucial. Doing so involves facing—and fixing—the big picture, not just dabbling with details as has been our habit. We would like the challenges we encounter to be handled in a democratic way, but in any case, they must be dealt with if we are to survive.

Ecoeconomics

Governments must begin to view economics in its broadest sense as well, as part of the planetary ecosystem, operating under the laws of evolution. Doing so would change the way we see ourselves relating to other things on our planet. The land we live on would then be given a value greater than that of the resources that can be extracted from it, and the welfare of future generations would be taken into account in government decisions.

In its 1990 Human Development Report, the United Nations Development Program suggested the creation of a "human development index" to measure the economic well-being of a country, combining GDP per capita with life expectancy, adult literacy rate, and purchasing power. Economist Herman Daly and theologian John Cobb have gone further and developed an "Index of Sustainable Economic Welfare," which takes a wide range of factors into account: air and water pollution, cropland losses, the consequences of automobile use, and income inequality.[8] By incorporating natural and human factors into a system like this, governments would better understand how their operations and decisions affect people and our planet.

Planning for and Protecting the Future

National governments have great impact on our future. When we make changes in government structures we must do so in ways that work to protect that future. Forecasts must be open-ended (i.e., they must include all available data that can affect them) and combine all trends and probabilities, so that more realistic, honest predictions can be made. Robert Ornstein and Paul Ehrlich contend that "Governments should institutionalize the long-term view.... The United States Government might establish a 'foresight' institute, which (like the present National Science Foundation) is relatively insulated from political interference."[9] Such an institute could overlook all of the functions of government to note whether they are acting in a coherent way and dealing with the world holistically.

SOLVING PROBLEMS

We must develop a method of applying a broad spectrum of human talents and abilities to solve an overwhelming number of difficult problems, and not rely so heavily on politicians pressed by selfish interests. This will be difficult to do in democratic systems, but we must find a way.

Governments not only need to reduce the time that elapses between the discovery of a problem and its solution, they must also devise a way to search for problems before they make themselves known to us. They must also find ways to expand the number of problems that can be dealt with at one time, or we will fall ever further behind.

By getting to the bottom of problems—to their causes—we can find real and lasting solutions, which could solve other problems in the process. Successfully addressing the fundamental reasons people take drugs, for example, would also reduce crime and the spread of AIDS, improve discipline in the classroom, and save our government much money. By looking at the wide range of effects caused by problems and their possible solutions, we can maximize the positive outcomes of our solutions and minimize negative consequences. We could also solve more problems effectively at less cost.

A government agency that would, without interference, search out as-yet-undetected problems and look into the now-ignored interactions between government agencies and programs could perform an invaluable service. It would draw attention to problems before they become serious and save billions of dollars now spent on programs and agencies that work against each other. Besides investigating current problems, this agency could direct Congress to the causes of problems it wishes to deal with and make them, and the public, aware of side effects of legislation under consideration. This agency would operate similarly to the General Accounting Office and would report its findings to both the legislative and administrative branches of government as well as to the public.

While the adversarial approach now prevalent within legislatures and between branches of government does serve some useful purposes, it often makes sensible problem-solving difficult. The posturing involved can be so distracting that it produces distorted, if not grotesque, results. We must tone down this method of resolving issues and where better techniques can be found, use them.

RESTRUCTURING GOVERNMENTS

A government's structure affects what it and its society do, and how they work—smoothly or inefficiently. When government works poorly it can even pervert well-intended undertakings.

Changes can be made in governmental structure to increase the likelihood of solving problems effectively, working more harmoniously with nature, and ensuring a decent environment for future generations. Robert C. North explains what is needed: "The fundamental problem is how to build . . . balancing mechanisms into societies in ways which will be flexible, largely self-regulating, and as efficiently, equitably, and democratically administered as possible."[10]

One of the most effective means of influencing behavior is self-regulating feedback loops. Democracy, capitalism, currency systems, bookkeeping, studies of social conditions, and environmental impact statements form negative feedback loops.* When working well, they can direct and coordinate activities to minimize negative side effects.

Social systems such as governments are very complex and have convoluted feedback loops. The behavior of most of them, however, can be changed by input applied at a small number of influence

*As noted earlier, negative feedback is where a portion of the output of a system or process is reintroduced into the system or process in order to correct output that has varied from the desired direction.

points. Unfortunately, public officials often do not know where these points are.[11] Studying such systems to learn where these points of influence are offers great potential for restructuring governments to better control the performance of these systems.

When the costs of polluting, now mostly borne by the public, are turned back on polluters, polluters will have good reason to control their emissions. If parties that extract resources from the earth paid society for the privilege of doing so instead of being subsidized in various ways, the costs of these resources would increase. People would then use them more wisely and have an incentive to share them with future generations. If drivers had to pay the real cost of driving (maintenance of streets and sewers, the cost of policing traffic, medical costs for those affected by pollution, etc.) there would be much less driving, with sizable savings to society, which would not have to pay these now hidden costs. In addition, our urban areas would be vastly improved. Although difficult, it is important and right that we find ways to turn the cost of damage done to the environment and future generations back onto the perpetrators.

Societal and economic structures need to be redesigned so that protecting the environment, conserving natural resources, and promoting peace and justice are rewarded, not penalized. Taxes can help with this by guiding human activities into constructive directions. Delayed inheritance taxes on family businesses and farms would help keep these in family ownership. Taxes on buying and selling stocks would decrease speculation and increase stockholders' interest in the operations of the companies they own because they would then look on them as long-term investments. Such taxes would also help pay for the many services governments provide to businesses.

Governments must replace their current chaotic, disjointed layers of organization and rules with arrangements that work as well-coordinated wholes that address the needs of their citizens. Damaging feedback loops must be replaced with constructive ones. Measures that encourage conservation should supplant those, such as the oil depletion allowance, that foster waste.

All these things need to be done as simply and straightforwardly as possible. Expanding populations, resource shortages, and increasing vulnerability to many dangers squeeze ever more tightly. In response, we have adapted burdensome regulations and controls that kill creativity and much of the satisfaction we find in our work. In architecture, for example, the complex sets of regulations that govern building construction consume huge amounts of designers' time and often leave little room for innovation, except in superficial elements such as ornamentation and color. We must find ways to manage to live safely in a complicated world without destroying the joy of living.

Restructuring government and its operations can not only make it possible to address what is now an unmanageable number of problems, but also make it the natural thing to do.

IMPROVING CAPITALISM

Every nation has some form of economic system—barter, capitalism, or socialism, for example. In recent years the market economy (capitalism) has proven itself far superior to centrally controlled systems (socialism). Nevertheless, capitalism has faults and more are appearing. These include dependence on a perpetually growing GDP to keep people employed; an increasing disparity between the rich and the poor; the tendency for economic power to be concentrated in fewer and fewer hands; and the separation of ownership and responsibility in corporations. These problems hurt individuals, communities, and the environment.

It has been suggested that technology, not foreign competition, is replacing good manufacturing jobs faster than new ones are created.[12] The service industry is replacing some of these jobs, but with poorly paid menial ones. Many people have been laid off not because they do their work poorly, but because their positions have been eliminated.

We are letting these things happen to us—standing in awe of capitalism, and assuming that in time it will correct our problems. It has not occurred to us that capitalism may not fulfill our dreams, or that we might do better by actively seeing to it that our economic system serves our needs, instead of our passively serving its needs.

The term "controlled capitalism" arose out of the Great Depression, an event which "proved" that capitalism needed controls to maintain a balance between boom and bust. Laws were enacted which have done much to stabilize the economy and moderate the excesses of unfettered financial activities. While many of those controls please conservative businesspeople, others do not please them (such as taxes that spread the wealth, worker protection legislation, and environmental regulations). These businesspeople try to convince the rest of us that we are better off without such laws. Rather than heeding their cries, we had best look at what history taught us about taming capitalism, and use this knowledge to benefit us all.

While the market does not provide employment to everyone who wants it, ironically, it leaves many things undone or done poorly. Our streets and parks are often dirty and in poor condition, our children are badly educated in large classes, and the government does not have funds to adequately enforce environmental, immigration, and health regulations. If we did not revere the market economy so greatly, we could take care of these things and in the process provide many new jobs. By giving everyone a meaningful place in society and paying them fairly, we would reduce unemployment, the need for social services, poverty, drug use, and crime—saving money on law enforcement and jails.

To help tame capitalism, groups such as nonprofit organizations and professional societies could set up committees to evaluate its success in providing for human needs in a democratic, environmentally sound way. Where failures, conflicts, or problems are found, these committees would notify their organizations and the public so that adjustments to the system could be made. By thoughtfully examining our economic machine, we would surely discover other things we could do to improve society.

CREATING INCENTIVES FOR ETHICAL BEHAVIOR

When we act as individuals, we manage to keep many of our nastier characteristics under control. Some of these controls disappear or are weakened when we act through our governments. We are more removed from the effects of our actions and often unaware of them. We do not see the results of eliminating school lunches for needy children, so we may be glad to accept the excuses for doing so. We are happy to pay fewer taxes. We bury our sense of obligation by being part of the crowd. People such as Idi Amin and ethnic leaders in the former Yugoslavia legitimize and bring out the worst in people by encouraging hatred and condoning atrocities such as "ethnic cleansing." We must devise ways to keep this from happening, and instead, keep our nobler instincts in control of not just ourselves but our governments.

Today, through the United Nations, many countries of the world profess to follow goals of establishing and maintaining peace and justice, helping poor nations, and protecting the environment. However, in practice nations fall far short of these professed intentions. Nations get by with hypocritical behavior because it is what we expect and because we tolerate it. It is how things are done! When people, the media, and most nations expect and demand straightforward behavior, national governments will be hard-pressed to behave differently.

As mentioned earlier, the evolutionary process and ecosystems depend on the workings of totally "selfish genes," the survival instinct of any species. In order to protect their gene pool, worker ants energetically perform a variety of tasks for the benefit of their colony. In human society, self-centered genes produce, among other things, some wonderful altruistic human beings, science, and art. If we consciously set out to do so, we should be able to restructure systems such as democracy and capitalism to produce more humane results even though the motivations of many of their citizens and functionaries are selfish. Providing better financial rewards and prestige for worthwhile work, for example, will provide a selfish motive for people to do it.

CIVIC RESPONSIBILITY

Of Citizens

Some people want to do their job well, treat others fairly, take good care of their families, and otherwise just be left alone. This is understandable, but there is a catch. A democracy can only be what its citizens make it. If they do not want to assume the responsibilities involved, such as being informed, participating in civic projects, and voting, in time they will have something else.

People must hold their leaders accountable. To do so, they need to know what those leaders are doing and planning. When politicians can only address people by wedging short, emotion-arousing sound bytes between the television entertainment in which people are more interested, democracy is not working properly. Nor is it working properly when leaders tell people only what they wish to hear.

There are two ways democracies might be strengthened. The first, the best—if it is possible—is to have everyone behave more responsibly. The second involves an unthinkable nonquestion: Should the right to govern be limited to people willing be responsible citizens?

Some people go to a lot of trouble to inform themselves, take their civil and personal responsibilities seriously, and contribute to the welfare of others. Others have no interest in public affairs and assume little responsibility for their family or even themselves. The votes of both types count the same. Although we urge everyone to vote, we do not ask them to be informed.

Perhaps the right to vote should be earned by demonstrating a certain level of political awareness and knowledge of current public concerns. This could pose risks, but the hazards of not doing so may be far greater. On the one hand, the least educated people might be taken advantage of. On the other, as problems become more complex and governments' responses more far-reaching, decisions become more critical and the consequences of bad decisions more dangerous. In the future we will not be able to bumble by unscathed as easily as we have in the past.

In democracies, it is essential that citizens understand the political process, the effect of special interest groups, and the motives and characteristics of the people who lead them. People need to see how their own involvement and demands encourage politicians to be honest or hypocritical and irresponsible. Emphasis on "game playing" in high school and university civics classes where students take on different roles could demonstrate how this works.

In order to understand issues we or our representatives can affect, we need to understand the milieu in which those issues occur. The media do a poor job of presenting this if they do it at all, but there is already an alternative to which we can turn. Over the past decades many excellent books have been written about particular nations and regions and the world. *The State of the World 1984* and later editions, by the Worldwatch Institute, explains environmental conditions on our planet. *China Wakes*, by Nicholas D. Kristof and Sheryl WuDunn, gives an overview of the people, politics, and economics of China today. *The Ends of the Earth*, by Robert D. Kaplan, describes dangerous and deteriorating situations in parts of the world we ordinarily know or care little about. Such books give one a far better understanding of places and conditions today than newspapers, television, or viewing countries through the windows of air-conditioned buses can.

Unfortunately many people do not read books. To help remedy this, foundations could fund, sponsor, and promote shorter versions of such books and encourage the writing of others. These texts would provide background information to readers so they gain a better understanding of the world they live in and can make informed decisions. We need to elevate the importance of background in the public mind. We must teach schoolchildren the importance of background so that they develop a habit of asking for information on things. The end result would be a public that responds more intelligently to our government's foreign policy.

I attended a progressive primary school connected to a teachers' college which had neither grades nor corporal punishment, and students had a voice in what they studied. Although we did not wave the flag or pledge allegiance to it, we gained a deep appreciation of democracy and learned how to be responsible. I think we would all do better if this feeling of responsibility and a clearer understanding of what democracy really is were widespread among our citizens.

Of Leaders

We must demand that our leaders also act responsibly. As our forefathers made personal sacrifices to gain and build democracy, we should expect our politicians to be willing to do the same in order to nurture and preserve it. Doing and saying whatever it takes to win the next election does not meet this criterion.

We cannot allow politicians to hide their real identity behind faces designed by pollsters. We must ask them to reveal their beliefs and intentions so that we are not hoodwinked, and thereby can vote intelligently. If candidates are not willing to do this, they are not suited to be in a government "by and for the people." It is up to us to see that they are not elected. To help with this, the media and the independent agency I suggested to present the pros and cons of issues to the public could keep a record and publicize politicians' promises and actions.

Today we need leaders who understand the dangerous problems we face and want to correct them. We need to get some of the best minds into the process of making important decisions. Jonas Salk was concerned about this: "What is needed are individuals who are consciously constructive rather than destructive to counterbalance the negative trends that now prevail, and to reverse them."[13]

Our present system of selection favors candidates who stage a good campaign, and people accept this as the way things are. This outlook must change. We must see to it that governments don't remain the playground of the vain and power-hungry.

I have an idea that will never sell, but it is not without merit: Plumbers, teachers, and nurses are required to take examinations and be certified to do their jobs. How about requiring people who run for government office to demonstrate their qualifications? Why not examine the candidates and let their potential employers, the voters, see how they measure up? Associations of economists, environmentalists, scientists, geographers, and historians could develop and administer tests. Besides testing them, this would give the candidates an incentive to learn some things they badly need to know.

The organization we need to maintain world order should do everything possible to remove proven dangerous individuals from positions of power in their nations before they become a menace to their own people and the world. Perhaps we should even require people who run for public office to take psychological tests.

The process of choosing people for important, nonelective government positions (cabinet ministers, department heads, and ambassadors, for example) can also be improved. Today they are chosen by people whose greatest strength may be winning public office, and the main criteria for selection may be political. This is good for politicians but bad for government. Businesses often rely on agencies to locate talent, and governments rely on specialized departments to hire workers. Ad hoc committees consisting of generalists backed by appropriate specialists could evaluate people being considered for appointment to important government positions. This would replace political connections as a criterion for selection with competence and suitability for the position to be filled.

THE CHALLENGE OF DEMOCRACY

Many of us think it inevitable that democracy will persist and spread across the planet. However, we have no good reason to believe this. As J. Arnold Toynbee has pointed out in his *Study of History*, the record of past civilizations does not provide reason for optimism.

There are many warning signs we should notice: Political candidates increasingly depend on well-financed special interests for support; qualified older workers on all levels are losing their jobs and cannot find satisfying new ones; employers find it difficult to obtain capable young workers with good work habits; drug abuse, crime, violence, and terrorism are increasing; growing numbers of children are brought up in dysfunctional "families"; pressures created by a swelling population and shrinking resource base are mounting; and most alarmingly, the public does not seem to care, unless it is in "their own back yard."

These problems tax the democratic process, and threaten the system itself. Without democracy, hope for solving the above and other difficulties is left to the pleasure of powerbrokers. Winston Churchill described democracy as being the least bad of all forms of government. But is it working well enough to deal with the greater number of problems it will face in the future? Many of them will be ones we have ignored.

Can a democracy survive when it is the "circle of followership" described in chapter 9, where everyone follows everyone else? We have gotten by until now, but can our form of government endure over time, especially as strains and stresses mount? Personally, I would like greater certainty.

Our forefathers fought to gain democracy, and if we wish to keep it we must live up to its challenges.

15

Making It Happen

I do not want the ideas in these pages to languish in remote dark corners of our divided minds or on dusty library shelves. I would like them to make a difference. Getting people to do what you believe they should, however, can be like standing on a bridge over a freeway shouting at the cars to stop. A plan for getting them to stop is needed. This chapter describes some tools and tasks that can be helpful, and suggests a strategy to implement the thoughts presented in the past two chapters.

WE MUST DEPEND ON OURSELVES

We cannot count on politicians and business leaders to initiate the changes needed to protect our planet and ensure a peaceful future. Most of them do not comprehend the seriousness of our present situation, nor do they look beyond the next election or the next several annual budgets. Winning primaries or increasing current profits forces them to focus on what is close, rather than on the big picture. They have only a superficial and spotty understanding of broader problems, and when they do voice concern for them, it is usually only to win our approval.

Politicians do listen to their constituents. However, the public is largely unaware of many of the problems we face and of what we should do about them. It wants its leaders to provide benefits, not worry about what it sees as abstract, irrelevant issues. To get leaders to behave more responsibly, we need a new breed of citizen: one more knowledgeable about himself and the problems humans face, people who think more clearly and who care. In his book *Earth in the Balance,* Al Gore (who understood how governments work) expressed the pressing need for us to learn and overcome our complacency: "Like it or not, we are now engaged in an epic battle to right the balance of our earth, and the tide of this battle will turn only when the majority of people in the world become sufficiently aroused by a shared sense of urgent danger to join an all-out effort."[1]

EDUCATE!

The key to engendering a sense of urgency lies in education. Today, most of us think of ourselves as "educated," well prepared for the world we live in, but in fact we are not. The education we have received is neither complete nor suited for the needs of today. It is geared to furthering individual survival in society and strengthening one nation as it competes against others, not protecting our species or our planet as a whole.

Once all creatures instinctively knew what they needed to know to survive and flourish—and they used this information. As the amount of knowledge available to humans grew, it became much more than was needed for survival, or than any human mind could absorb. This presented people with the need to pick and choose information. Scholars developed ideas of what an educated person should know, and this became accepted, traditional. As new information piled up, problems that people had never encountered before appeared, and people's ideas of what education should encompass fell behind. Our minds are now filled with information that has no bearing on our survival, and lack much that is essential for it.

We do not have a clear idea of what an education should include today. Tradition, conflicting ideas and preconceptions, the needs of business, and our subconscious drives push for one thing or another. An ever-changing set of theories developed by educators meant to undo the failures of past theories keeps sending teaching methods and subject matter off in new directions. This, plus what we are bombarded with by advertisers, entertainers, and our peers, makes up our current social education process. When I taught architecture I saw a curriculum changed much for the worse by current fashion and the haggling of a politically divided faculty unwilling to discuss or agree on anything sensible.

What to Teach?

We ought to start by asking what is needed today? The answer should not be provided by any one group—educators, humanists, scientists, or businesspeople, for example. It calls for input from society's most intelligent members of all different personality types and backgrounds as well as from less gifted people with direct personal experience in many areas.

We all must know what is essential to support and manage ourselves and our families. This includes a basic knowledge of our bodies, parenting, and finances. We also want to enjoy our life and explore its richness and possibilities, which means being introduced to our cultural heritage, the beauty of nature, and our own inner potential. Society has needs as well. It needs people with the wide range of abilities and knowledge required to carry out its many functions.

But taking care of ourselves, our families, and the everyday affairs of society is not enough. In a democracy, the citizens are responsible for how their country is governed. Thus they must have a broad education. While people in countries with other forms of governments may have no direct say in what their government does, they also need to be informed. Their concerns cannot be totally ignored, as their leaders and bureaucrats often come out of their midst. And, as in Spain in 1977 when Franco returned power to the royal family (which led to the return of democracy), and more recently in some countries emerging from communist rule, the people may gain political power.

In a democracy, responsible citizenship requires a basic understanding of how government works in theory and practice. We need to know not just the mechanics of how representatives are elected, as we are taught in school now, but also the parts played by special interest groups as well as their assistant, public apathy. Ideally, we should understand specific issues too so that our representatives are less influenced by selfish groups. But there is much more. Citizens of all nations need to know certain things about our planet and the natural laws that govern it. But we are not taught how our planet works! Everyone must learn at least the basics of this.*

Because we have an ever greater impact on what is happening to the world around us, we also need to understand ourselves. Understanding ourselves and others encourages reasonable behavior and strengthens democracy.

One of the primary aims of education has been, and should be, to help us step beyond our primitive instincts. Until recently in our evolution, generally we instinctively acted in the best interest of our species. But today, with the impact we have on our planet and the tremendous power we have at our disposal, our natural responses can be dangerous. Someone pursuing status purely for personal satisfaction or seeking political office merely to satisfy vanity or a need for power is acting much like an animal seeking food or sex. We must be taught to replace misdirected drives and goals with a conscious effort to protect our species from the many threats it faces.

We need to learn that what really produces happiness or a feeling of well-being depends more on our inner attitudes than on exterior circumstances. Until recently, there has been little research on happiness. Psychologists David G. Myers and Ed Diemer suggest that to overcome much of the current confusion about it, we must study happiness scientifically.[2] Our own experience and university research show that having more money or things does not make people more happy. Happiness increases as the distance between what we have and want, and between what our neighbors have and what we think we should have diminishes.[3] Directing our attention beyond ourselves and doing something for others increases happiness. I believe that happiness or a feeling of worth cannot be gained by direct pursuit—they result instead from involvement in beneficial activities.

*In order to foster some of this necessary learning, I will discuss this in some detail later in this chapter.

Mihaly Csikszentmihalyi, professor of psychology at the University of Chicago, uses the term "flow" to describe a state of mind in which one becomes immersed in a task that fully challenges one's skills and abilities.[4] Achieving "flow" on a regular basis brings satisfaction, joy, and a sense of well-being into one's life. Unfortunately, too often we choose work for reasons of money, prestige, or social acceptance, and businesses believe that their obligation to their employees ends with material rewards. We should be taught to pursue what brings us real meaning, and also that employers should feel an obligation to provide jobs that are personally rewarding to their employees.

By helping us reorient our default concepts of what is desirable. teachers can help us lead more rewarding lives, and by doing so, help society change its goals. People who pursue ends that bring them true contentment and happiness will waste fewer resources, cause less environmental damage, and be less likely to cause other people harm than those who seek expensive pleasures, goods, or power.

We need to be made aware of our brain's weaknesses and limitations, so that we are less arrogant about our abilities and ideas and more open to learning the truth. Although our ability to reason is limited, we must be trained to think more clearly and taught how to resist having our thoughts manipulated.

Likewise, we need to understand how we interact in groups of all kinds. We should be made aware of our irrational surrender to peer pressure and fads, how we can get caught up in the hysteria of a crowd and relinquish our personal integrity to a larger entity so that we can learn how to avoid falling into "groupthink."

Our education and culture must stress the shallowness of current materialistic status symbols and urge people to respect and evaluate each other on personal qualities rather than on wealth, power, or fame.

Our minds are now increasingly filled with trivia, and for most people, the "Information Highway" may only be a way of adding to this. Trivia now crowds out much of what we should know. I am not sure how we can change this, but we must find ways to get people to devote a larger part of their thinking to matters of consequence.

The Teachers

To bring about changes in what we are taught, we must look to our teachers—families, religious organizations, interest groups, peers, schools, the media, and advertisers. Of these, the easiest to work with and influence are the media, schools, and some churches.

THE MEDIA

The news media are the fastest way to inform citizens about matters of public concern. They reach more people than books and magazines do. They are a critical component in the feedback loop linking public responses to issues with government action. However, they usually provide information in disconnected bytes and operate with serious shortcomings. Despite their limitations, they can do considerably better at awakening the public to the various threats to our survival.

The news industry is increasingly owned by business conglomerates which have one purpose: profit. Journalism concentrates on current happenings and is undertaken by people with personality types attracted by crisis and conflict. Sensationalism sells to a public clamoring for excitement and entertainment. The public thinks that what is important is limited largely to what is printed or shown on a television screen. This makes it difficult to arouse people to issues that are not regularly covered by the media, such as the fundamental causes of problems and the slow global transformations that are threatening us today. This is the reality that those who would change reporting must recognize and work with. Fortunately, there is light in this bleak picture. Some media owners, journalists, and members of the public do want high quality, responsible journalism.

To achieve this, the media must face up to the fact that they are the major and sometimes sole means of informing adults of current conditions, events, and new discoveries. They therefore must provide people with what they need to know to be responsible citizens of their community, nation, and world. This means seeing that citizens are informed on topics where their influence is crucial. Journalists also have the task of making it clear that humanity's future depends on the well-being of people as well as a healthy planet, and they should be made to feel embarrassed when they forego this important responsibility in order to satisfy selfish goals.

Too often journalists confine themselves to the easy task of reporting what people say without evaluating it. This sounds objective, but it can confuse us. Zealots are misleading the public today on critical issues by their statements in the media. Some tell us that there is no environmental crisis, that evolution is a hoax, or the world's population growth will stop as people in the underdeveloped nations become prosperous. Evidence, if publicly considered, would discredit such claims. When discussing an endangered species, an ecologist who has objectively studied a forest should be considered more credible than a spokesperson for a lumber company. A scientist recognized for her achievements should be considered more trustworthy than someone who makes money telling businesspeople that global warming is a conspiracy. Journalists should not routinely portray people presenting conflicting statements as equally credible. Instead, they should present facts, such as the qualifications, backgrounds, and affiliations of those presenting ideas, in order to help people discern the truth. Journalists should also point out whether these individuals represent the beliefs held by the majority of members of the discipline involved, or of a small minority.

Reporters need to be awakened to the importance of environmental problems, long-term concerns, and slow change. They must be given a holistic view of the world and made to realize that everything in it is in some way related to other things. As I will discuss in a broader context later in this chapter, environmental and peace groups and foundations could help by conducting and funding conferences, short courses, and scholarships, and by encouraging the introduction of new courses at journalism schools.

Techniques for presenting serious issues in formats that appeal to a wider public would encourage and help the news media to cover these subjects. I suspect that many people now confined to devoting their talent to selling beer and soap would like to help make matters of crucial concern interesting to us all. If the media would, for example, in an interesting way, provide us with more information about successful efforts to reduce adolescent crime and improve conditions in developing countries, people would be more willing to have their governments support such projects. An example of the potential of the media to arouse public awareness was provided by the coverage of the three whales caught in the Arctic ice described in the introduction to this book. Although in other years, people knew or cared little about large numbers of whales that died this way, in 1988, with the help of journalists, the world was awakened to the plight of these helpless creatures and did what they could to help them. We should examine instances like this to learn what we can do to better stimulate people's concern for the serious problems needing their attention.

Newspapers, news magazines, and television should produce daily or weekly reports on environmental happenings just as they do on sports, entertainment, and financial matters. One way to keep the public aware of major long-term trends would be to have monthly or six-week reviews of them in our sources of current events. These reviews would evaluate progress, point out worsening conditions, and remind citizens when situations have not changed. *Vital Signs* and *State of the World*, published and updated annually by the Worldwatch Institute,[5] and "Living on Earth," produced by National Public Radio, provide the kind of information that is needed. But this knowledge should not be limited to those who look for it. The media has a responsibility to bring it to the attention of all citizens, especially those who make political or economic decisions. Some print media, including the *New York Times* and *Time* magazine (which occasionally publish issues such as "Our Precious Planet" devoted to special topics) fulfill this obligation. Unfortunately, as can be observed in the routine reporting of these publications, their own journalists seem to be unaffected by the stories they cover.

Primary and High Schools

The education of young people should be an integrated whole, not a conglomeration of neatly packaged separate subjects or units. It should include (if it does not already), among other things, some knowledge of science, mathematics (all young people need to understand the principle of exponential progression and how it affects populations), geography, sociology, history, literature, personal finance, citizenship, conflict resolution, clear thinking, parenting, and for those who can handle it, a second language. Specific topics to be touched on in these subjects would include the basic principles of the scientific method, ecology, evolution, thermodynamics, general systems theory, political science, economics, psychology, philosophy, and propaganda techniques. How all of these interact with each other must also be discussed. Students also need to understand the value of both doers and thinkers and of how society benefits by being made up of people with different personality types.

This sounds like a lot. However, these subjects need not be taught as they are now—filled with much that satisfies tradition and the educational bureaucracy, but which does little for students or society. Important principles of many so-called complex things can be effectively taught to even young children.

Kenneth Boulding has said that economics should be broadly taught to everyone, and that the main part could be boiled down to one-half hour. My sixth-grade teacher spent about two weeks on some of the techniques used by advertisers and political propagandists. This had such an impact that today I recognize just what advertisers are doing to me, and when I shop, wherever possible, foolish as it may be, I avoid aggressively advertised items. I think that we could, with great benefit, also teach our children how to better avoid being drawn in by charismatic fanatics.

What students really need is to be able to understand and get along with themselves, others, and nature. Emphasis could be placed on the understanding of concepts and principles, leaving details to those who later specialize. A special effort must be made to obtain science, mathematics, and logic teachers who have an aptitude, understanding, and love for these subjects. Rote learning must be minimized and students must be encouraged to think for themselves. A careful paring off of material not needed to meet our goals will make room for that which is essential.

A comparison of American education with that of Canada, Europe, Russia, China, and Japan shows that we can expect much more from our students. By adapting itself to the varying intellectual capacities of students and expanding opportunities for each student to do their best, teaching can bring students toward their highest potential. Much room can be made for meaningful learning by reducing the cerebral clutter that inundates our lives today.

Lessons and Projects

Young people need to know a number of important facts about our minds, particularly that our system of perception is limited. For example, children need to learn that the world is not the same as the one we think we see; that we have trouble connecting ideas in our divided minds; and that our strong emotions tend to predominate over our weak ability to reason. It has been suggested that museums be built to demonstrate these facts to visitors.

Student participation in evaluating systems and model building introduced in schools at the proper times would demonstrate how the world works. Such projects can be simple, but as noted above, can teach important and lasting lessons. This would be an excellent way to help students gain a basic understanding of the scientific method, our close connection to nature, and the vulnerabilities of modern society. It would give them practice in observing slow change, finding the root causes of problems, holistic thinking, weighing future needs of society, and setting goals and priorities.

For example, students could compare the overall effects of a family living in an inner city neighborhood on nature with those of a family living in a distant suburb. What are the effects on air pollution, of time and money spent on travel? Young people could learn about where hamburgers come from: The importance of the microbes in the soil for producing the grain that is fed to beef cattle, how much land is needed to raise the grain and graze the cattle for one hamburger, and how this affects the economy and environment in developing countries. They could consider how we plan for a future that has enough food for everyone as the world's population increases and from this, determine what the options are.

Students could spend some time looking for and at nonquestions (that we often take for granted) and try to develop the habit of finding and delving into them. They should also learn how to find information located in different parts of our brains, then make connections. What is the relationship, for example, between the claimed need for economic growth and water pollution?

Some projects for students include discussing different problems and nonquestions and deciding which are most important to correct: relieving traffic by building a new highway, safely storing nuclear waste, building a school swimming pool, or stopping soil erosion? What will happen if economic growth continues indefinitely in order to provide jobs? Should it be stopped? When? In recent times, what have countries gained or lost by starting a war? How should exhaustible resources be used over time, and what will people do in the future when they are used up? How should the world's wealth be divided? What will the future of the world's people be like, and what factors today affect it? Is there a universal ethic? How are we dependent on natural systems? Specially trained and talented teachers could go from classroom to classroom and lead these discussions.

Because we have traveled so far away from nature in our synthetic world, we need to make a concerted effort to reestablish contact with her. Primitive camps—within one hundred miles of campers' homes so that they learn about their own surroundings—without running water or electricity, where campers (guided by knowledgeable, enthusiastic staff) provide the daily necessities of life, would help. Children could stay a week—or better, a month. It should be made fashionable for adults to have this experience at least once as well.

Teachers should practice what they preach—example is an important teaching tool for children. Example can be used to teach clear thinking, avoiding stereotypes, looking for causes, observing slow change, planning, saving, and energy conservation. David Orr of the Department of Environmental Science, Oberlin College, suggests that students "examine resource flows on campus: Food, energy, water, materials, and waste. Faculty and students should together study the wells, mines, farms, feedlots, and forests that supply the campus as well as the dumps where you send your waste."[6]

In the day-to-day relationships between teachers and students, the teaching of ethical values by example cannot be avoided.[7] Teachers must recognize this and ground their behavior on the basic set of ethics agreed upon by world moral leaders. Teachers should also work to change students' role models. Instead of the football hero, the cheerleader, or the most attractive student, role models should be those who distinguish themselves academically or do the most to help others. Rather than rock stars, role models should be people who have truly contributed to improving the human condition.

Besides being taught how democracy works, students should be taught about its difficulties and the dangers that come when it fails.

They should learn about their own responsibility to help maintain it, and about our natural tendency to abuse it. They could play roles in a game that demonstrates what happens when people act in their own interest instead of caring about the welfare of everyone. Learning these things is better insurance for preserving democracy than distorted patriotic versions of history that leave one disturbed when the truth is learned later in life.

Ideas such as that it is honorable to die for one's country, in any instance, or that the United States has always behaved nobly should be replaced by teaching genuine patriotism, which includes respect for all people and all life.

The United Nations' Educational, Scientific, and Cultural Organization (UNESCO) should establish basic minimum requirements for teaching unbiased history and about environmental problems and conditions in developing countries. It should also investigate school systems around the world and draw international attention to those that teach ethnocentrism or any form of hate or bigotry. Teaching that one's own people are superior to others, or that certain lands elsewhere rightfully belong to one's own country would indicate that the nations involved may be a source of future problems. Such tenets should be a matter of international concern and borne in mind by nations when establishing foreign and trade policies.

We cannot expect instant success in achieving the goals described above. We still must work with bureaucracies; there are limited numbers of good, caring teachers; and the public is stingy when it comes to education. However, as people gradually learn that our real worst enemies lie within human nature and the structure of our institutions, they will shift resources (including human talent) from war, extravagance, and waste to educating and improving the lives of people.

Instead of taking almost anyone willing to accept the low esteem and rewards we give teachers, for example, we have to get some of our best people into the classroom. People who have a feel for science are particularly needed to instill a respect for the scientific method in all students. Early or late retired people who have experienced life and achieved something would have a lot to give our young people. We could seek out and find exceptional retiring people and offer them a chance to teach a high school class for one or more years. I think there would be many who would like to do this. What could be a more worthwhile activity and investment in our future than providing our young people with the most inspiring teachers we could find?

College and Adult Education

There are possibilities for major improvement in higher education as well. In colleges and universities, all students should be required to take a set of integrated general courses designed to show the interconnectedness of all things and our dependence on our planet and each other. These courses would examine subjects like science, economics, and history in more detail than in primary schools. This common background of university and college students would ease communication among them in later life. Our colleges and universities should serve as examples as well, practicing energy conservation, avoiding chemicals on lawns, constructing beautiful buildings, and investing endowments and assets in ethical enterprises.

Nor need we neglect adults who are no longer attending schools. Night, correspondence, television, and even Internet courses on clear thinking and other subjects that further human survival can be offered and promoted. Clear thinking can be sold as a means for earning more money, getting a better job, and improving family and personal relationships. Public television could present programs where influential people—writers, academics, and clerics—would be challenged by panels of critical thinkers to rationally defend their positions.

Reward can also be an effective tool. I know of no award equivalent to the Nobel Prize that is bestowed on individuals or organizations for their work in protecting our future. Recognition given to individuals who further planetary survival in areas such as science, advocacy, government, education, and business, for example, would draw public attention to environmental concerns. I believe there are savvy, well-endowed philanthropists who might fund such a project.

Finally, a computer program we might call Earth Game could be developed and sold. In an entertaining way it would demonstrate how, by interacting with natural systems and each other, different policies and actions affect our planet and lives. It would allow players to assume different roles, and in pursuing their interests and trying to win, see what would happen if, say, more automobiles were manufactured and sold in underdeveloped counties, international funding for birth control were eliminated, or international trade doubled. Such computer programs have been developed, but to my knowledge, not as tools of educational recreation.

Religious Organizations

Religions have the potential to do much good. Their leaders provide ideas and inspiration to ready audiences seeking moral guidance. Christianity, for example, calls for peace, charity, and justice toward all humans, in addition to good stewardship of the earth; other religions have similar teachings. Secular groups promoting various aspects of human survival should reach out to religions and support their efforts to further these ends. Groups promoting human survival should work with religious organizations to show them how to put their ideals into practice and to overcome the preoccupation with overly rigid dogma, rules, and formalities that sometimes afflicts them.

Progressive clerics can help by teaching human survival to be a matter of ethics. Our effect on the entire future of the human race (and whether humans even survive into that future) is something all caring people are surely concerned with.

SURVIVAL STUDIES

We do not yet know all of the things we should do in order to protect our future. New dangers are constantly coming to light. Until recently we were unaware of the threat posed to animal and human fertility by the chemicals we are spreading over the earth. We may not even be aware of the greatest perils we now face. We must discover them as quickly as possible.

In 1984 John Herz proposed a new subdiscipline of international studies to be entitled "Survival Research": "If global survival, that is surviving the threat of nuclear extinction and the threat to mankind posed by the combined effects of our population explosion, exhaustion of resources, and destruction of the environment, is the central problem of our age, I submit that it is the duty of political scientists—

as of all social scientists—to think about these issues and search for solutions."[8] The article including this assessment was submitted to the journal *Political Science* and rejected. Apparently the editors saw no merit in these ideas. Nor do most of today's political science practitioners seem interested in this problem as they go about examining relative trivialities in great detail. Nevertheless, this proposal makes sense and could be a keystone of our pursuit of survival.

The discipline Herz suggests should be open ended, examine everything that relates to human survival, and include input from all fields of study. It could be called "Survival Studies" to include both identification of problems and creative proposals for solutions. Survival Studies would be an extension of ecology, concentrating on the relationship among the human mind, society, and the earth's biosphere. We must understand the effect of our lifestyle on the world around us and learn how our minds, government, business, religion, and our evolutionary development interact. Excellent work has already been done in this direction in subjects such as general systems theory, general living systems theory, and sociobiology.[9] This trailblazing needs to be coordinated and expanded.

An important aspect of this new discipline would be the study of information and thinking in the world today: the "noosphere" of Teilhard de Chardin.* It would examine how we determine what data is to be sought out, how information is stored, integrated, retrieved, and how it moves us into action. A better understanding of this would help us greatly.

Survival Studies would also evaluate how existing political and economic systems interact with each other, and what parts they and potential alternatives could best play in the world system. It would propose new ways of governance. Ideal models, such as those described in chapter 14, designed to better meet human needs within planetary capabilities, would be devised and evaluated.

---

*". . . outside and above the biosphere, . . . an added planetary layer, an envelope of thinking substance, to which, for the sake of convenience and symmetry, I have given the name of the Noosphere." Pierre Teilhard de Chardin, *The Future of Man* (New York: Harper & Row, 1969), p. 137.

DANGEROUS THINKING

Educating and learning are not enough. We will have to take more aggressive measures, too. Advocates of damaging thoughts should not be given the respect we now give them. These people are impeding efforts to protect our planet, control our numbers, and establish peace through enforceable world law. Others leave the subject of human survival to "God's will." Some groups resort to or condone assassinations or terrorism. Irrational groups are growing in number and through their fanaticism and single-mindedness are having a growing influence on our lives and political events. This must be taken seriously before it gets even worse.

Fighting these groups issue by issue or event by event is an unending battle that is never won and may even increase the resolve of their supporters and sympathizers. An important task for our thinkers, professionals, and other motivated individuals should be to challenge muddled thinking. The proponents of harmful thoughts should be asked difficult questions and be forced to defend their positions. Their beliefs should be confronted with scientific fact, and the contradictions within them exposed.

The papacy and others who oppose artificial birth control should be asked to explain what should be done when human numbers increase to a point where there is no possible way to prevent mass starvation. Groups that insist "creationism" be taught along with evolution as an equally valid viewpoint should be challenged to present their arguments beside those of notable scientists to bodies of people qualified to judge. Concerned citizens should not allow these people to stack school boards, as they now do, and then behave as they please. People and organizations that promote unending expansion must be forced to explain how this can be perpetuated indefinitely, and if not, when it will be stopped. Today, people are allowed to make illogical proposals without answering the obvious questions. We must force them to provide answers.

STRATEGY

Being aware of the problems we face, understanding what to do about them, and knowing the tools we can use to do so is not enough. We still have to *do* it, and that requires a plan, a strategy. Criticism has no value unless it produces positive results, so some suggestions as to how we can implement action are needed here, naive though they may be. My intent in the following is simply to stimulate further thinking by those more suited to the task.

There is only one place action can begin: with people who acknowledge the problems and want to do something about them. Many people see the dangers posed by pollution, habitat loss, or the proliferation of dangerous weapons. They join organizations, contribute to causes, write letters, and protest. They fight each battle as it comes, but every battle that is won is followed by others. Some of these fighters feel we are making progress, others are more pessimistic.

The Spark

A smaller number of people see the larger picture and recognize that we are losing the war. They know what this means and it is horrifying. These people are burdened with a terrible responsibility—getting the rest of us to join the fight. If they do not, no one else will, and we will go right on losing. There are some aware, committed people, but they are often isolated or have little power or influence. It is hard for them to dedicate themselves to something when they have little support and do not know what to do. Nevertheless, they can and must be the spark that lights the fire.

They can take heart from the example of zealous religious and political believers totally committed to their beliefs, some of whom have even died for their ideals. Their commitment gives them an influence that reaches far beyond their numbers. Those committed to thinking clearly should be able to do even better.

Those who are ready to light the fire may find others who share their concerns within organizations to which they already belong, or within their professions. By working together, they can expand their network and establish contact with others in groups that may have different immediate objectives, but who are concerned about the common war.

Over time, alliances should be made with notable individuals, scientists and other professionals within disciplines dealing with various aspects of current problems: agronomists, economists, and meteorologists, for example. People such as physicists Albert Einstein, Leo Szilard, and Henry W. Kendall, agriculturist Lester Brown, biologists Rachel Carson and Paul Ehrlich, writer Norman Cousins, and politicians Gaylord Nelson, Morris Udall, Bruce Babbitt, and Al Gore have dedicated not just their time but their reputations to the cause of human survival. Various movements concerned with survival are far stronger because of their work.

The Flicker

How much stronger the message would be if many prestigious individuals and professional organizations would—loudly, clearly, and in unison—take a public stand together and lend their support to shared concerns. This would persuade many people now bewildered by the many conflicting voices they hear. It would lessen the influence of hired public relations people and the proponents of "we are doing just fine and need more growth." A less confused, more concerned public would have an immediate impact on politicians who are now overwhelmed by special interest groups.

Actually, over the years there has been considerable activity along these lines. For example, journals such as *The Bulletin of Atomic Scientists* have been published; conferences attended by scientists and other thinkers have been organized; stands like the "World Scientists' Call for Action" at the Kyoto Climate Summit signed by over 1,500 scientists, including 104 Nobel laureates, have been taken and publicized. However, these efforts have not been as effective as they should have been at clarifying issues for the public or at arousing it to action. Ways must be found to give consequential statements greater impact, and to teach journalists the importance of recognizing the vast difference between public relations and serious science—so that the latter can be given the attention it deserves.

Most scientists and thinkers are aware of many threats to our well-being, and of the magnitude of these threats. However, for various reasons—other demands on their attention, apathy, or powerlessness—they do nothing meaningful about them. Our prestigious group of concerned individuals and organizations should confront them. They should point out how Kitty Genovese was slowly murdered as her neighbors stood by and watched. They should remind these scientists and thinkers how many Germans who saw the danger of Hitler were silent as he took over their country, and then it was too late. The rapid destruction of our planet's life support system is far more serious than anything Hitler could have done. Knowledgeable bystanders, if necessary, should be shamed into taking a stand.

Once a certain level of concern and involvement has been reached, it will gather power as people willingly, even eagerly, participate. The public thus aroused can be the kindling that helps the fire spread. Professional societies of scientists, engineers, and others should make emphatic public statements and take out full-page advertisements in newspapers and magazines where they can clarify hazy issues such as climate change. Although this has already been done in some cases, it would be more effective and convincing if it were a more unified effort carried out in a broader and more consistent way.

Our prestigious group of thinkers along with professional groups could launch the interdisciplinary Survival Studies and suggest projects for foundations to sponsor. They could pressure universities to teach all students about matters of survival.

These groups and organizations should set up conferences, seminars, and short courses aimed at educating political and civic leaders, journalists, teachers, businesspeople, and the general public. The Aspen Institute and the Vienna Academy have run programs along these lines. A conference for journalists and publishers, for example, could address the responsibility of media regarding the threats to humanity and our planet.

Organizations with active, concerned members, such as the Sierra Club, can design and promote short courses on subjects relating to human survival to be taught in elementary schools. In Cincinnati, for example, the Center for Peace Education has successfully promoted and conducted a conflict resolution training program in many local schools. It includes instructing teachers and setting up student mediator programs. Besides learning how to resolve disputes, students gain an understanding of what causes conflicts and why we often do so poorly at resolving them.

The ultimate aim of such projects would be to make the public aware of our current predicament, and enable it to see not just what can be done but the obstacles that hinder us from doing so. Everyone must learn about pressing issues, but more importantly must be given a fundamental understanding that leads to more lasting results.

❂ ❂ ❂

This is a good place to end our discussion.

If there is one thing we could benefit from, it is to observe and think more clearly, and to recognize our weaknesses in these areas. We need to see wholes instead of just disorderly accumulations of parts, and we need to listen and search for truth, instead of expounding misconceptions. If we did that—began to think more clearly—a lot of our problems would begin to be resolved.

Conclusion

Our brains are proving to be a double-edged sword. On the one hand, they have brought us many wonderful things, such as bread, cars, airplanes, penicillin, air-conditioners, and literature. On the other hand, they have given us weapons with which we can destroy each other many times over and have started a process that is rapidly eroding the earth's ability to support life. These brains, which evolved to help us survive in a far simpler world, have done very well when focussing on specific tasks. However, they are unable to deal safely with the total picture of what has followed. This totality has passed beyond our control, and if allowed to continue will lead to widespread social chaos, wars, mass starvation, and inconceivable human misery.

Perhaps it is inevitable that evolution eventually self-destructs or reaches a limit by producing a creature that has evolved far enough to make clever inventions, but not far enough to control and use them wisely. This leaves a dangerous time gap. Biological evolution is far too slow to introduce the necessary levels of intelligence, accountability, and altruism for humans to span this gap rapidly enough. Fortunately, there is another possibility.

Evolution takes place not only in plants and animals, but in society as well. This means we can do what no living thing has ever done before. We can become conscious as a species, actively decide to protect our environment, and live at peace with each other. We can replace unsatisfying, selfish, damage-causing goals driven by ancient instinctive urges with those that produce true contentment and happiness. No species has previously needed to act consciously as none—until very recent times—has thrown the world's ecology off balance. Being conscious as a species is a challenge that we have not yet recognized or accepted. It means acknowledging common problems and goals and, together, working on them intelligently.

Over time our concerns have gradually extended from ourselves and our offspring to our family, tribe, and nation-state. We must now take a giant step and extend our concerns to all humanity and life; not just for the present, but on into the long-term future as well.

There are three ways we can move into that future. First, we can unconditionally embrace the seeming benefits of modern technology and medicine and leave future generations to cope with the consequences. Second, we can abandon modern technology and medicine, resubjecting ourselves to the laws of nature and allowing them to take their course in reducing our population in unpleasant ways, thus restoring the ecological balance. Or, third, we can accept our responsibility to establish stability between nature and the new powers we have unleashed, and replace war with law as a means for settling disputes. We can select one of these alternatives by conscious choice, or by default—which is what we are now doing.

There is reason to believe that we can make the responsible choice. We have shown that we can rise above our limitations and act on a higher level. People have made radical changes in the past; they have altered their religious beliefs, created democracies and developed new economic systems. They have eradicated slavery and established widely accepted standards for human rights. Changes in beliefs cleared the way for ideas that made science, industry, and the availability of capital for investment possible. We are able to override the selfish genes that are not concerned about the future or with acting altruistically. No other species has ever been able to do this.

But the question remains, will we make it? Some people feel we must make optimistic predictions so that we do not become discouraged. I think it is better to admit honestly that we do not know. This helps avoid making people complacent when their involvement in needed change is essential. People must see that we are on a course headed toward disaster—but that there is hope—if they participate.

The quality of our environment has badly deteriorated from what it once was, and continues to do so. The sooner and more resolutely we act, the better chance we have of countering these changes, and the less damaged and overpopulated the planet we leave our descendants will be. Some people will point out that technology is bound to make life better. But I do not believe that a 9,000-channel, full-wall, 3-D television will compensate for 10 billion people having to depend on inadequate food supplies.

We can do all of the things we need to do to solve our problems. It would be a terrible thing if we did not try. Unfortunately, right now the burden of trying rests on very few shoulders, on those who understand our situation and have some idea of what needs to be done. We cannot blame others for doing nothing. It is people like you and me who must get things moving. So let's get to work!

Afterword

I would like to hear from people who have comments or ideas on how to implement change—or who would like to participate in bringing it about. I do not promise to answer all replies individually, but in time, you will hear from me. Perhaps we can even do something. You can reach me by:

Mail: 944 Lenox Place
Cincinnati, Ohio 45229

Fax: 513-751-6662

Email: pseidel@fuse.net

Notes

INTRODUCTION

1. Tom Rose, *Freeing the Whales: How the Media Created the World's Greatest Non-Event* (New York: Carol Publishing Group, 1989).

2. Ludwig von Bertalanffy, *General Systems Theory* (New York: George Braziller, 1968), p. 8.

3. Jonas Salk, *Anatomy of Reality: Merging of Intuition and Reason* (New York: Columbia University Press, 1983), p. 122.

4. Doris Lessing, *Prisons We Choose to Live Inside* (New York: Harper & Row, 1987), p. 5.

5. F. E. Trainer, *Abandon Affluence and Growth* (n.p.: Zed Books, 1965). Quoted in *World Watch* (September–October 1990): 8.

6. The Environmental Pollution Panel of the President's Science Advisory Committee, *Restoring the Quality of Our Environment* (Washington, D.C.: The White House, November 1965), pp. 111–33.

7. Fairfield Osborn, *Our Plundered Planet* (Boston: Little, Brown and Company, 1948).

8. William Vogt, *Road to Survival* (New York: William Sloan Associates, 1948), quote appears on dust jacket of the original edition.

PART I: OUR ANCIENT BRAIN

1. Albert Szent-Gyorgi, "The Persistence of the Caveman," *Saturday Review* (July 7, 1962): 11.

1. THE WORLD AS WE PERCEIVE IT

1. Robert Ornstein and Paul Erlich, *New World New Mind* (New York: Doubleday, 1989), p. 73.
2. Elizabeth F. Loftus, *Eyewitness Testimony* (Cambridge, Mass.: Harvard University Press, 1979).
3. Ornstein and Erlich, *New World New Mind*, pp. 74–75.
4. Paul Erlich, *The Population Bomb* (New York: Ballantine Books, 1968), p. 4.
5. Patricia Goldman-Rakie, "Working Memory and the Mind," *Scientific American* 262, no. 3 (September 1992): 112.
6. Edward O. Wilson, *In Search of Nature* (Washington, D.C.: Island Press, 1996), p. 172.

2. THE LIMITATIONS OF OUR BRAIN

1. Ornstein and Erlich, *New World New Mind*, p. 101.
2. Daniel Goleman, *Vital Lies, Simple Truths* (New York: Simon & Schuster Touchstone Book, 1985), p. 65.
3. Jeremy Campbell, *The Improbable Machine* (New York: Simon & Schuster, 1989), p. 108.
4. Allen R. Kahn and Berry C. Deer, "Limits of Human Information Processing: Supplementation with Machine Intelligence" (unpublished paper, 1989), p. 4.
5. Michael S. Gazzaniga, *The Social Brain* (New York: Basic Books, 1985), p. 110.
6. Michael S. Gazzaniga, *Nature's Mind* (New York: Basic Books, 1992), p. 95.
7. Marvin Minsky, *The Society of the Mind* (New York: Simon & Schuster, 1986), pp. 66–67 and 224. Minsky is a computer scientist. See also the work of neuroscientist Gazzaniga, particularly *The Social Brain*, pp. 84–91 and 117.

8. Gazzaniga, *The Social Brain,* p. x.

9. Robert Ornstein, *Multimind* (Boston: Houghton Mifflin, 1986), pp. 72–73.

10. Minsky, *The Society of the Mind,* p. 17.

11. Ornstein, *Multimind,* p. 23.

12. Minsky, *The Society of the Mind,* p. 40.

13. Kenneth E. Boulding, *The Image* (Ann Arbor: University of Michigan Press, 1956), p. 111.

14. Al Gore, *Earth in the Balance* (Boston: Houghton Mifflin, 1992), p. 239.

15. Campbell, *The Improbable Machine,* pp. 14–15.

16. Gustave Le Bon, *The Crowd* (*La Psychologie des Foules,* Paris, 1895; New York: Viking Press, 1960), p. 64.

17. Jean Piaget, *The Construction of Reality in the Child* (New York: Basic Books, 1954).

18. Massimo Piattelli-Palmarini, *Inevitable Illusions* (New York: John Wiley & Sons, 1994).

3. THOSE EVER-COMPELLING PRIMARY DRIVES

1. Nicholas Humphrey, *A History of the Mind* (New York: Simon & Schuster, 1992), pp. 101–14.

2. Goleman, *Vital Lies, Simple Truths,* p. 244.

3. Ervin Staub, *The Roots of Evil* (Cambridge: Cambridge University Press, 1989), p. 239.

4. Mark Green et al., *There He Goes Again* (New York: Pantheon Books, 1983), p. 99.

5. Gore, *Earth in the Balance,* p. 220.

6. Goleman, *Vital Lies, Simple Truths,* p. 241.

7. Elting E. Morison, *Men, Machines, and Modern Times* (Cambridge, Mass.: MIT Press, 1966), chap. 6.

8. Eric Fromm, *Escape from Freedom* (New York: Holt, Rinehart & Winston, 1941; Avon Library, 1964), p. 225.

9. Ibid., p. 216.

10. Ruth Benedict, *Patterns of Culture* (New York: Pelican, 1946; Mentor, 1948), pp. 168–205.

11. A series of interesting experiments came up with similar results.

Henri Tajfel, social psychologist at Bristol University in England, and his colleagues demonstrated that one can predictably change a person's behavior by assigning him to a certain group. Without having prior knowledge of the assigned group or its purpose, nor knowing any of its members, he will immediately stand up for them against the members of another. Nigel Calder, *The Human Conspiracy* (New York: Viking Press, 1976).

12. Andrew Bard Schmookler, *Out of Weakness* (New York: Bantam Books, 1988), p. 12.

13. Robert Jay Lifton, *The Nazi Doctors* (New York: Basic Books, 1986), pp. 4 and 5.

14. Le Bon, *The Crowd,* p. 26.

15. Craig Haney, Curtis Banks, and Philip Zimbardo, "Interpersonal Dynamics in a Simulated Prison," *International Journal of Criminology and Psychology* 6 (1968): 279–80.

16. Zygmunt Bauman, *Modernity and the Holocaust* (Ithaca, N.Y.: Cornell University Press, 1989), p. 168.

17. Vilmos Csányi, *Evolutionary Systems and Society: A General Theory of Life, Mind, and Culture* (Durham, N.C.: Duke University Press, 1989), pp. 175–78.

18. Stanley Milgram, *Obedience to Authority* (New York: Harper & Row, 1974), pp. 135–47.

19. Christopher R. Browning, *Ordinary Men* (New York: HarperCollins Aaron Asher Books, 1992).

20. Lifton, *The Nazi Doctors,* pp. 419–29.

21. Ibid., pp. 442–47.

22. Staub, *The Roots of Evil,* p. 241.

23. George Orwell, "England Your England," *A Collection of Essays* (San Diego: Harcourt Brace Jovanovich, 1993), p. 252.

4. WHEN WE COME TOGETHER

1. C. G. Jung, *Psychological Types* (Princeton, N.J.: Princeton University Press, 1973).

2. Walter Lowen, *Dichotomies of the Mind* (New York: John Wiley & Sons, 1982).

3. Jose Ortega y Gasset, *La Rebellion de las Masas* (1930; *The Revolt of the Masses,* New York: W. W. Norton, 1932), p. 44.

4. Jeffrey van Davis, *Norman Mailer: The Sanction to Write* (self-produced video documentary, 1982).

5. Ervin Laszlo, "Resistance to Innovation in Complex Systems: Application to Contemporary Society" (unpublished paper, ca. 1989), pp. 8–9.

5. THE PSYCHOLOGY OF SOCIETY

1. Friedrich Nietzsche, quoted in Goleman, *Vital Lies, Simple Truths,* p. 161.

2. Le Bon, *The Crowd,* p. 27.

3. Arthur Koestler, *Janus* (London: Hutchinson & Co., 1978; New York: Vintage, 1979), p. 95.

4. Richard Dawkins, *The Selfish Gene* (Oxford: Oxford University Press, 1976), pp. 206 and 208.

5. Aaron Lynch, *Thought Contagion* (New York: Basic Books, 1996), pp. 23 and 51.

6. Orrin E. Klapp, *Overload and Boredom* (New York: Greenwood Press, 1986), pp. 31.

7. David G. Myers, *The Pursuit of Happiness* (New York: William Morrow & Co., 1992), p. 43.

8. Klapp, *Overload and Boredom,* p. 14.

9. Otto Fenichel "On the Psychology of Boredom," in *The Collected Papers of Otto Fenichel* (New York: W. W. Norton, 1953), pp. 270–81.

10. Thomas Whiteside, *The Investigation of Ralph Nader* (New York: Arbor House, 1972).

11. B. Latané and J. M. Darley, *The Unresponsive Bystander: Why Doesn't He Help?* (New York: Appleton-Century-Crofts, 1970).

12. Anthony Storr, *Human Destructiveness* (New York: Ballantine Books, 1992), p. 124.

6. OVERLOAD AND OTHER DILEMMAS

1. Ortega y Gasset, *La Rebellion de las Masas,* p. 90.

2. Bill Moyers, *A World of Ideas* (New York: Doubleday, 1989), p. 182.

3. G. W. F. Hegel, from the Introduction, *The Philosophy of History* (1837).

4. Laurie Garrett, *The Coming Plague* (New York: Farrar, Straus & Giroux, 1994), chap. 16.

5. Ornstein and Erlich, *New World New Mind,* p. 248.

6. Herman E. Daly and John B. Cobb Jr., *For the Common Good* (Boston: Beacon Press, 1989), p. 34.

7. John H. Herz, "The Responsibilities of a Political Scientist in an Age of Threatened Survival" (published remarks from the Tenth Annual CUNY Political Science Conference, New York, December 7, 1984), p. 2.

8. Ortega y Gasset, *La Rebellion de las Masas,* p. 112.

7. BELIEFS

1. Campbell, *The Improbable Machine,* p. 233.

2. Gazzaniga, *The Social Brain,* pp. 139 and 146.

3. James E. Alcock, "The Belief Engine," *Skeptical Inquirer* (May/June 1994): 17.

4. Ervin Laszlo, *The Inner Limits of Mankind* (London: One World Publications, Ltd., 1989), pp. 35–36.

5. Ornstein and Erlich, *New World New Mind,* p. 82.

6. Alcock, "The Belief Engine," p. 15.

7. Gazzaniga, *The Social Brain,* p. 139.

8. Schmookler, *Out of Weakness,* p. 284.

9. Gore, *Earth in the Balance,* p. 181.

10. Thomas Berry, *The Dream of the Earth* (San Francisco: Sierra Club Books, 1988), p. 37.

11. Ascribed to Chief Seattle in 1854 (in a U.S. National Park Service exhibit in Seward, Alaska).

12. Alfonzo and Margaret Ortiz, eds., *To Carry Forth in the Vine* (New York: Columbia University Press, 1978).

13. Carolyn Merchant, *The Death of Nature* (New York: Harper & Row, 1980), p. 247.

14. Berry, *The Dream of the Earth,* p. 80.

15. Merchant, *The Death of Nature,* p. 214.

16. Berry, *The Dream of the Earth,* p. 41.

17. Merchant, *The Death of Nature,* p. 185.

18. Andrew Bard Schmookler, "All Consuming: Materialistic Values and Human Needs," *Center Review* (Spring 1990): 15.

19. Dawkins, *The Selfish Gene,* chap. 12.

20. Ornstein, *Multimind,* p. 188.

8. ETHICS AND VALUES

1. Unknown American Indian chief, *How Can One Sell the Air?* (Summertown, Tenn.: Book Publishing Co., 1988).
2. Marjorie Hope and James Young, "Thomas Berry and a New Creation Story," *The Christian Century* (August 16–23, 1989): 750.
3. Fromm, *Escape from Freedom,* p. 150.
4. Ibid., pp. 276–77.
5. Myers, *The Pursuit of Happiness.* See also "Science Pursues Happiness," special report, *The Futurist* (September–October 1997).
6. See R. H. Tawney, *Religion and the Rise of Capitalism* (Glouster, Mass.: P. Smith, 1962).
7. Quoted in Louis E. Boone, *Quotable Business* (New York: Random House, 1992), p. 198.
8. Paul Hawken, *The Ecology of Commerce: A Declaration of Sustainability* (New York: HarperCollins, 1993), pp. 99–100.
9. "Global Epidemic—Getting Worse," *Infact Update* (Summer 1994), p. 5, also see "Tobacco Imperialism," *Multinational Monitor* (January–February 1992).
10. "Global Epidemic—Getting Worse."
11. Geoffrey C. Bible, in the Philip Morris Companies, Inc., Annual Report, p. 4.
12. Amy L. Domini with Peter D. Kinder, *Ethical Investing* (Reading, Mass.: Addison-Wesley, 1984).
13. Quoted in Staub, *The Roots of Evil,* p. 263.
14. Kenneth Boulding, *Ecodynamics* (Beverly Hills, Calif.: Sage Publications, 1978), pp. 319–20.
15. The Food and Agricultural Organization of the United Nations, *The State of Food and Agriculture* (Rome, 1992), p. 22.
16. Ervin Laszlo, *The Choice: Evolution or Extinction?* (New York: Tarcher/Putnam, 1994), pp. 46 and 200.
17. Ibid., p. 27. See also Dominique Lapierre, *The City of Joy* (Garden City, N.Y.: Doubleday, 1986), for a remarkable story of impoverished families in Calcutta, India.
18. Nicholas D. Kristof and Sheryl WuDunn, *China Wakes* (New York: Random House, 1994), p. 390.
19. Ibid., p. 15.

9. LEADERSHIP AND FOLLOWERSHIP

1. Based on data collected and furnished by Common Cause, a nonprofit, nonpartisan citizens' lobbying organization promoting open, honest, and accountable government based in Washington, D.C.

2. Herbert A. Simon, *Administrative Behavior,* 3d ed. (New York: Free Press, 1976), pp. xxix and xxx.

3. Le Bon, *The Crowd,* pp. 130 and 132.

4. Ervin Laszlo, *Bifurcation* (unpublished manuscript, May 1989), p. 176.

5. Kenneth Boulding, *Evolutionary Economics* (Beverly Hills, Calif.: Sage Publications, 1981), p. 173.

6. C. A. Gibb, "Leadership," in *Handbook of Social Psychology,* 2d ed., Vol. 4, G. Lindzey and E. Aronson, eds. (Reading, Mass.: Addison-Wesley, 1969), p. 218.

7. Halvor Egtun, *Aftenposten* (Oslo: March 28, 1995, reprinted in translation in *World Press Review* [July 1995]: 7).

8. Ortega y Gasset, *La Rebellion de las Masas,* p. 18.

9. Ibid., pp. 14–15.

10. Cohen and Solomon, " 'Crime Times' News Exploits Fears," *Liberal Opinion Week* (Creators Syndicate, Inc., June 27, 1994).

11. See Ervin Laszlo, *The Age of Bifurcation* (Philadelphia: Gordon and Breach, 1991).

12. Robert C. North, *Survival and Other Futures* (unpublished manuscript, 1988), p. 399.

13. Ibid., p. 397.

14. Gary Wills, *Certain Trumpets: The Nature of Leadership* (New York: Simon & Schuster, 1994).

15. Gore, *Earth in the Balance,* p. 305.

16. Marcus Felson, *Crime and Everyday Life* (Thousand Oaks, Calif.: Pine Forge Press, 1944).

10. THE WORKINGS OF GOVERNMENT

1. David Malin Roodman, "Paying the Piper," *Worldwatch Paper* 133, December 1996 (Washington, D.C.: Worldwatch Institute), p. 6.

2. Hawken, *The Ecology of Commerce,* ref. to Russell Sabin, "Study Outlines Cost of Smoking in State," *San Francisco Chronicle* (August 1, 1992).

3. Based on figures of Stephen B. Goddard, "The Driving Costs of Transportation," *St. Louis Post-Dispatch* (July 8, 1994).

4. Nicholas Pigou, *A Study in Public Finance* (London: Macmillan & Co., Ltd., 1928), p. 99. Discussed in Hawken, *The Ecology of Commerce*, p. 82.

5. Lester R. Brown, President of Worldwatch Institute, personal correspondence with Peter Seidel.

6. Dawkins, *The Selfish Gene*, pp. 117–18.

7. "Teen Moms Cost U.S. $34 Billion," *The Cincinnati Post* (June 2, 1994): 1A.

8. Bill McKibben, *New York Times* (May 5, 1997), op-ed page.

9. H. Allen Smith, *The Complete Practical Joker* (New York: William Morrow & Co., 1959), pp. 129–30.

10. Julian Borger, "Back Inside the Dead Zone," *The Guardian* (London, quoted in *World Press Review*, August 1994: 11–12).

11. Jörg Albrecht et al., *Die Zeit* (Hamburg, November 11, 1994, translated and presented as "Russia's Total Mess," in *World Press Review*, February 1995: 10–11).

12. Garrett, *The Coming Plague*, p. 617.

13. Marvin J. Centron with Owen Davies, "The Face of Future Terrorism," *The Futurist* 28, no. 6 (November–December 1994): 10.

14. L. F. Ivanhoe, "Get Ready for Another Oil Shock," *The Futurist* 31, no. 1 (January–February 1997): 20–23.

15. Ibid.

16. John H. Herz, *International Politics in the Atomic Age* (New York: Columbia University Press, 1959), p. 214(n).

11. ORGANIZED VIOLENCE

1. Kristof and WuDunn, *China Wakes*, p. 73.

2. Ruth Leger Sivard, *World Military and Social Expenditures* (Washington, D.C.: World Priorities, 1996), p. 8.

3. Storr, *Human Destructiveness*, p. 21.

4. Michael Renner, "Budgeting for Disarmament," *Worldwatch Paper* 122, November 1994 (Washington, D.C.: Worldwatch Institute).

5. Quoted from Nicholas Mosley, *Julian Grenfell, His Life and the Times of His Death, 1888–1915* (London: Weidenfeld & Nicholson, 1976), p. 239.

6. John H. Herz, *The Nation-State and the Crisis of World Politics* (New York: David McKay Co., 1976), pp. 9–10.

7. Andrew Bard Schmookler, *The Parable of the Tribes: The Problem of Power in Social Evolution* (Boston: Houghton Mifflin, 1984), pp. 21–23.

8. Robert C. North, *The World That Could Be* (New York: W. W. Norton, 1976), pp. 120–21.

12. BUSINESS AND ITS SIDEKICK

1. Gaylord Nelson, National Press Club luncheon address, Washington, D.C., April 20, 1990.

2. Herz, *International Politics in the Atomic Age,* p. 276.

3. Ross Gelbspan, *The Heat Is On* (Reading Mass.: Addison-Wesley, 1997), p. 5.

4. Ibid., pp. 3–4.

5. Derived from figures taken from the Bureau of the Census, *Statistical Abstract of the United States,* 1995–96. Published as *The American Almanac* (Austin, Tex.: The Reference Press).

13. WHAT TO DO?

1. Salk, *Anatomy of Reality,* p. 113.

2. Laszlo, *The Inner Limits of Mankind,* p. 28.

3. Robert Ornstein, *The Roots of the Self* (San Francisco: Harper San Francisco, 1993), p. 180.

4. Gazzaniga, *Nature's Mind,* p. 137.

5. Kenneth E. Boulding, *Human Betterment* (Beverly Hills, Calif.: Sage Publications, 1985), p. 55.

6. Lessing, *Prisons We Choose to Live Inside,* p. 60.

7. Laszlo, *The Inner Limits of Mankind,* pp. 50–51.

8. Aaron Lynch, *Thought Contagion* (New York: Basic Books, 1996), p. 176.

9. Cited in Goleman, *Vital Lies, Simple Truths,* p. 239.

10. Berry, *The Dream of the Earth,* p. 117.

11. The General Evolution Research Group, "The Evolution of Cognitive Maps," Position Paper, Simposio Internazionale del General Evolution Research Group (Bologna, Italy, May 16, 1989), p. 7.

12. Laszlo, *The Inner Limits of Mankind,* p. 104.

13. Paul Kurtz, *Forbidden Fruit: The Ethics of Humanism* (Amherst, N.Y.: Prometheus Books, 1988), p. 18.

14. Berry, *The Dream of the Earth,* p. 184.

15. Victor Hugo, *The Alps and Pyrenees,* trans., John Mason (London: Bliss Sands & Co., 1898).

16. Roger Sperry, *Science and Moral Priority* (New York: Columbia University Press, 1983).

17. Ibid., pp. 125–26.

18. Boulding, *Evolutionary Economics,* p. 154.

19. Herz, *The Nation-State,* p. 301.

20. John Herz, "Some Observations on Engaging in 'Survival Research,' " unpublished paper, p. 11.

21. Dawkins, *The Selfish Gene,* p. 224.

22. Lessing, *Prisons We Choose to Live Inside,* p. 19.

14. GOVERNMENTS MUST HELP

1. North, *The World That Could Be,* pp. 59–60.

2. Dwight D. Eisenhower, quoted in Lawrence Abbott, *World Federalism—What? Why? How?* (Arlington, Va.: World Federalist Association, n.d.).

3. John F. Kennedy, quoted in ibid.

4. George Bush, quoted in ibid.

5. Albert Einstein, quoted in ibid.

6. Herz, *International Politics in the Atomic Age,* p. 32.

7. General Omar Bradley, speech at the dedication of a memorial statue, Philadelphia, Pennsylvania, August 10, 1952.

8. Herman E. Daly and John B. Cobb Jr., *For the Common Good* (Boston: Beacon Press, 1989), pp. 401–55.

9. Ornstein and Ehrlich, *New World New Mind,* p. 249.

10. North, *The World That Could Be,* p. 108.

11. Jay W. Forrester, "Counterintiutive Behavior of Social Systems," *ZPG National Reporter* (June 1971).

12. Jeremy Rifkin, "Civil Society in the Information Age," *The Nation* (February 26, 1996), p. 11.

13. Salk, *Anatomy of Reality,* p. 118.

15. MAKING IT HAPPEN

1. Gore, *Earth in the Balance,* p. 269.

2. David G. Myers and Ed Diemer, "The Science of Happiness," *The Futurist* (September–October 1997), Special Report, p. 2.

3. Myers, *The Pursuit of Happiness,* chap. 2 and 3. See also Alden E. Wessman and David F. Ricks, *Mood and Personality* (New York: Holt, Rinehart & Winston, 1996), pp. 170–73.

4. Mihaly Csikszentmihalyi, *Finding Flow* (New York: Basic Books, 1977), pp. 28–34.

5. Worldwatch Institute, 1776 Massachusetts Avenue NW, Washington, DC 20036.

6. David Orr, "What Is Education For? *Annals of Earth* 8, no. 2 (1990): 13.

7. Lawrence Kohlberg, *Essays on Moral Development,* vol. 1 (San Francisco: Harper & Row, 1981), p. 1.

8. Herz, "The Responsibilities of a Political Scientist."

9. See Ervin Laszlo, *The Systems View of the World* (New York: George Braziller, 1972); Jay W. Forrester, *World Dynamics* (Cambridge, Mass.: Wright Allen Press, 1971); James Grier Miller, *Living Systems* (New York: McGraw-Hill, 1978); and Edward O. Wilson, *Sociobiology, The Abridged Edition* (Cambridge, Mass.: Harvard University Press, 1975). See also Boulding, *The Image,* regarding the rationale for the establishment of a new science, "eiconics," based on the concept of the image.

Bibliography

Albrecht, Jörg, Patricia Fuller, Dirk Kufbjuweit, and Walter Suller. *Die Zeit* (Hamburg, November 11, 1994). Translated and presented as "Russia's Total Mess," in *World Press Review*, February 1995): 10–11.

Alcock, James E. "The Belief Engine," *Skeptical Inquirer* (May–June 1995): 17.

Bartecchi, Carl E., Thomas D. MacKenzie, and Robert W. Schrier. "The Global Tobacco Epidemic," *Scientific American* 272, no. 5 (May 1995): 46.

Bauman, Zygmunt. *Modernity and the Holocaust.* Ithaca, N.Y.: Cornell University Press, 1989.

Benedict, Ruth. *Patterns of Culture.* New York: Pelican, 1946; Mentor, 1948.

Bernstein, Shawn. *The Rise of Air Pollution Control as a National Political Issue: A Study of Issue Development*, paper, the Graduate School of Arts & Science, Columbia University, spring, 1982.

Berry, Thomas. *The Dream of the Earth.* San Francisco: Sierra Club Books, 1988.

Boone, Louis E. *Quotable Business*. New York: Random House, 1992.

Borger, Julian. "Back Inside the Dead Zone," *The Guardian.* Quoted in *World Press Review* (August 1994): 11–12.

Boulding, Kenneth E. *Ecodynamics.* Beverly Hills, Calif.: Sage Publications, 1978.

Boulding, Kenneth E. *Evolutionary Economics.* Beverly Hills, Calif.: Sage Publications, 1981.

———. *Human Betterment.* Beverly Hills, Calif.: Sage Publications, 1985.

———. *The Image.* Ann Arbor: University of Michigan Press, 1956.

Browning, Christopher R. *Ordinary Men: Reserve Police Battalion 101 and the Final Solution in Poland.* New York: HarperCollins, 1992.

Calder, Nigel. *The Human Conspiracy.* New York: Viking Press, 1976.

Campbell, Jeremy. *The Improbable Machine.* New York: Simon & Schuster, 1989.

Centron, Marvin J., with Owen Davies. "The Face of Future Terrorism," *The Futurist* 28, no. 6 (November–December 1994): 10.

Cohen and Solomon, "'Crime Times' News Exploits Fears," *Liberal Opinion Week* (June 27, 1994).

Colander, D., and A. Kalmer. "The Making of an Economist," *Economic Perspectives* 1 (1987): 95–111.

Csányi, Vilmos. *Evolutionary Systems & Society: A General Theory of Life, Mind, & Culture.* Durham, N.C.: Duke University Press, 1989.

Csikszentmihalyi, Mihaly. *Finding Flow.* New York: Basic Books, 1997.

Daly, Herman E., and John B. Cobb Jr. *For the Common Good.* Boston: Beacon Press, 1989.

Dawkins, Richard. *The Selfish Gene.* New York: Oxford University Press, 1989.

Egtun, Halvor. *Aftenposten.* Oslo: March 28, 1995. Reprinted in translation in *World Press Review* (July 1995): 7.

Environmental Pollution Panel of the President's Science Advisory Committee, The. *Restoring the Quality of Our Environment.* Washington, D.C.: The White House, November 1965.

Erlich, Paul. *The Population Bomb.* New York: Ballantine, 1968.

Ehrlich, Paul, and Anne Ehrlich. *Betrayal of Science.* Washington, D.C.: Island Press, 1996.

Felson, Marcus. *Crime and Everyday Life.* Thousand Oaks, Calif.: Pine Forge Press, 1944.

Fenichel, Otto. *The Collected Papers of Otto Fenichel.* New York: W. W. Norton, 1953.

Food and Agricultural Organization of the United Nations, The. *The State of Food and Agriculture.* Rome, 1992.

Forrester, Jay W. "Counterintuitive Behavior of Social Systems," *ZPG National Reporter* (June 1971).

Fromm, Eric. *Escape from Freedom.* New York, Holt, Rinehart & Winston, 1941; Avon Library, 1964.

Garrett, Laurie. *The Coming Plague.* New York: Farrar, Straus & Giroux, 1994.

Gazzaniga, Michael S. *Nature's Mind.* New York: Basic Books, 1992.

———. *The Social Brain.* New York: Basic Books, 1985.

Gelbspan, Ross. *The Heat Is On.* Reading, Mass.: Addison-Wesley, 1997.

General Evolution Research Group. *The Evolution of Cognitive Maps.* Position paper, Simposio Internazionale del General Evolution Research Group, Bologna, Italy, May 16, 1989.

Gibb, C. A. "Leadership." In G. Lindzey and E. Aronson, eds., *Handbook of Social Psychology*, 2nd ed., Vol. 4. Reading, Mass.: Addison-Wesley, 1969.

"Global Epidemic—Getting Worse," *Infact Update* (Summer 1994).

Goddard, Stephen B. "The Driving Costs of Transportation," *St. Louis Post-Dispatch*, July 8, 1994.

Goldman-Rakie, Patricia. "Working Memory and the Mind," *Scientific American* 262, no. 3 (September 1992): 112.

Goleman, Daniel. *Vital Lies Simple Truths.* New York: Simon & Schuster Touchstone Book, 1985.

Gore, Al. *Earth in the Balance.* Boston: Houghton Mifflin, 1992.

Green, Mark, et al. *There He Goes Again.* New York: Pantheon Books, 1983.

Haney, Craig, Curtis Banks, and Philip Zimbardo. "Interpersonal Dynamics in a Simulated Prison," *International Journal of Criminology and Psychology* 6 (1968): 279–80.

Hawken, Paul. *The Ecology of Commerce: A Declaration of Sustainability.* New York: HarperCollins, 1993.

Herz, John H. *International Politics in the Atomic Age.* New York: Columbia University Press, 1959.

———. *The Nation-State and the Crisis of World Politics.* New York: David McKay Co., 1976.

———. "The Responsibilities of a Political Scientist in an Age of Threatened Survival." Published remarks from the Tenth Annual CUNY Political Science Conference, New York, December 7, 1984.

———. "Some Observations on Engaging in 'Survival Research,' " unpublished paper.

Hope, Marjorie, and James Young. "Thomas Berry and a New Creation Story," *The Christian Century* (August 16–23, 1989).

Humphrey, Nicholas. *A History of the Mind.* New York: Simon & Schuster, 1992.

Ivanhoe, L. F. "Get Ready for Another Oil Shock," *The Futurist* 31, no. 1 (January–February 1997): 20–23.

Jung, C. G. *Psychological Types.* Princeton, N.J.: Princeton University Press, 1973.

Kahn, Allen R., and Berry C. Deer. "Limits of Human Information Processing: Supplementation with Machine Intelligence," unpublished paper, 1989.

Kaplan, Robert D. *The End of the Earth.* New York: Random House, 1996.

Klapp, Orrin E. *Overload and Boredom.* New York: Greenwood Press, 1986.

Koestler, Arthur. *Janus: A Summing Up.* London: Hutchinson & Co., 1978; New York: Vintage, 1979.

Kristof, Nicholas D., and Sheryl WuDunn, *China Wakes: The Struggle for the Soul of a Rising Power.* New York: Random House, 1994.

Kulper, David. "David Ross Brower," *The Progressive* (May 1994): 38.

Kurtz, Paul. *Forbidden Fruit: The Ethics of Humanism.* Amherst, N.Y.: Prometheus Books, 1988.

Laszlo, Ervin. *The Age of Bifurcation.* Philadelphia: Gordon and Breach, 1991.

———. *Bifurcation.* Unpublished manuscript, May 1989.

———. *The Choice: Evolution or Extinction?* New York: Tarcher/Putnam, 1994.

———. *The Inner Limits of Mankind.* London: Oneworld Publications, Ltd., 1989.

———. "Resistance to Innovation in Complex Systems: Application to Contemporary Society." Unpublished paper.

Latané, B., and J. M. Darley. *The Unresponsive Bystander: Why Doesn't He Help?* New York: Appleton-Century-Crofts, 1970.

Le Bon, Gustave. *The Crowd* (*La Psychologie des Foules*). Paris, 1895; New York: Viking Press, 1960.

Lessing, Doris. *Prisons We Choose to Live Inside.* New York: Harper & Row, 1987.

Lifton, Robert Jay. *The Nazi Doctors.* New York: Basic Books, 1986.

Loftus, Elizabeth F. *Eyewitness Testimony.* Cambridge, Mass.: Harvard University Press, 1979.

Lowen, Walter. *Dichotomies of the Mind.* New York: John Wiley & Sons, 1982.

Lynch, Aaron. *Thought Contagion.* New York: Basic Books, 1996.

McCarthy, Joseph L. "Mr. Rogers' Neighborhood," *Chief Executive* (January–February 1996).

Meadows, Donella H., et al. *The Limits of Growth.* New York: New American Library, 1972.

Merchant, Carolyn. *The Death of Nature.* New York: Harper & Row, 1980.

Milgram, Stanley. *Obedience to Authority.* New York: Harper & Row, 1974.

Minsky, Marvin. *The Society of the Mind.* New York: Simon & Schuster, 1986.

Morison, Elting E. *Men, Machines, and Modern Times.* Cambridge, Mass.: MIT Press, 1966.

Mosley, Nicholas. *Julian Grenfell, His Life and the Times of His Death, 1888–1915*. New York: Holt, Rinehart & Winston, 1976.
Moyers, Bill. *A World of Ideas*. New York: Doubleday, 1989.
Myers, David G. *The Pursuit of Happiness*. New York: William Morrow & Co., 1992.
Myers, David G., and Ed Diemer. "The Science of Happiness," *The Futurist* (September–October 1997).
North, Robert C. *Survival and Other Futures* (unpublished manuscript, 1988).
———. *The World That Could Be*. New York: W. W. Norton, 1976.
Ornstein, Robert. *Multimind*. Boston: Houghton Mifflin, 1986.
———. *The Roots of the Self*. San Francisco: Harper San Francisco, 1993.
Ornstein, Robert, and Paul Ehrlich. *New World New Mind*. New York: Doubleday, 1989.
Ortega y Gasset, Jose. *La Rebellion de las Masas* (*The Revolt of the Masses*) (1930; New York: W. W. Norton, 1932.
Ortiz, Alfonzo, and Margaret Ortiz, eds. *To Carry Forth in the Vine*. New York: Columbia University Press, 1978.
Osborn, Fairfield. *Our Plundered Planet*. Boston: Little, Brown, & Co., 1948.
Piaget, Jean. *The Construction of Reality in the Child*. New York: Basic Books, 1954.
Piattelli-Palmarini, Massimo. *Inevitable Illusions*. New York: John Wiley & Sons, 1994.
Pigou, Nicholas. *A Study in Public Finance*. London: Macmillan & Co., Ltd., 1928.
Raiz, Dennis V., ed. *The Human Predicament*. Amherst, N.Y.: Prometheus Books, 1996.
Renner, Michael. "Budgeting for Disarmament," *Worldwatch Paper* 122, November 1994, Washington, D.C.: Worldwatch Institute.
Rifkin, Jeremy. "Civil Society in the Information Age," *The Nation* (February 26, 1996).
Rolston, Holmes, III. *Environmental Ethics*. Philadelphia: Temple University Press, 1988.
Roodman, David Malin. "Paying the Piper," *Worldwatch Paper* 133, December 1996, Washington, D.C.: Worldwatch Institute.
Rose, Tom. *Freeing The Whales: How the Media Created the World's Greatest Non-Event*. New York: Carol Publishing Group, 1989.
Ryan, John C. "Life Support: Conserving Biological Diversity," *Worldwatch Paper* 108, April 1992, Washington, D.C.: Worldwatch Institute.
Sabin, Russell. "Study Outlines Cost of Smoking in State," *San Francisco Chronicle*, August 1, 1992.

Salk, Jonas. *Anatomy of Reality*. New York: Columbia University Press, 1983.

Schmookler, Andrew Bard. "All Consuming: Materialistic Values and Human Needs," *Center Review* (Spring 1990).

———. *Out of Weakness*. New York: Bantam Books, 1988.

———. *The Parable of the Tribes: The Problem of Power in Social Evolution.* Boston: Houghton Mifflin, 1984.

Schumacher, E. F. *Small Is Beautiful.* New York: Harper & Row, 1973.

Seidel, Peter. "Cities in a Real World," *World Futures* 39, no. 4 (1994): 183–86.

———. "The Cost of Wealthy Modern Cities," *Indian Journal of Applied Economics* (January 1998) (special issue).

Simon, Herbert A. *Administrative Behavior,* 3d ed. New York: Free Press, 1976).

Sivard, Ruth Leger. *World Military and Social Expenditures.* Washington, D.C.: World Priorities, 1996.

Smith, H. Allen. *The Complete Practical Joker*. New York: William Morrow & Co., 1950.

Sperry, Roger. *Science and Moral Priority*. New York: Columbia University Press, 1983.

Staub, Ervin. *The Roots of Evil.* Cambridge: Cambridge University Press, 1989.

Storr, Anthony. *Human Destructiveness.* London: Grove Weidenfeld, 1991; New York: Ballantine Books, 1992.

Tawney, R. H. *Religion and the Rise of Capitalism.* Glouster, Mass.: P. Smith, 1962.

Terresa, Vincent, with Thomas C. Renner. *My Life in the Mafia.* Greenwich, Conn.: Fawcett Publications, 1974.

"Tobacco Imperialism," *Multinational Monitor* (January–February 1992).

Toynbee, J. Arnold. *A Study of History,* abridgement of vols. I–X by D. C. Somerwell. New York: Oxford University Press, 1947–1987.

Trainer, F. E. *Abandon Affluence and Growth.* N.p.: Zed Books, 1965. Quoted in *World Watch* (September–October 1990): 8.

Unknown American Indian chief, An. *How Can One Sell The Air?* Summertown, Tenn.: Book Publishing Co., 1988.

van Davis, Jeffrey. *Norman Mailer: The Sanction to Write.* Self-produced video documentary, 1982.

Vogt, William. *Road to Survival.* New York: William Sloan Associates, 1948.

von Bertalanffy, Ludwig. *General Systems Theory.* New York, George Braziller, 1968.

Wessman, Alden E., and David F. Ricks. *Mood and Personality.* New York: Holt, Rinehart & Winston, 1996.

Whiteside, Thomas. *The Investigation of Ralph Nader.* New York: Arbor House, 1972.

Wills, Gary. *Certain Trumpets: The Nature of Leadership.* New York: Simon & Schuster, 1994.

Wilson, Edward O. *In Search of Nature.* Washington, D.C.: Island Press, 1996.

———. *Sociobiology, The Abridged Edition.* Cambridge, Mass.: Belknap Press, 1975.

Worldwatch Institute, *State of the World 1984 to 1998.* New York: W. W. Norton & Co., 1984– .

Acknowledgments

It is interesting to look back and think of all those who helped make this book possible. I simply do not know all of them. As with the ideas the book encompasses and the disciplines that were necessary to bring it into being, they go far back in time.

I was lucky to have supportive parents who cheerfully tolerated an unending stream of questions. This was important. Friends and teachers, notably Alfred Caldwell, A. W. Kappus, Alfred Madson (my grammar school manual training teacher), Ludwig Mies van der Rohe, and Walter Peterhans have all influenced my way of looking at things. L. Hilberseimer, who used the Socratic method to teach planning at Illinois Institute of Technology, convinced me that all dwellings should receive direct sunshine in winter and that cars be put in their place—he suggested Lake Michigan.

As I mentioned in the introduction, in 1957 I read a book by Harrison Brown that planted a seed by discussing a variety of environmental and resource problems. In the late 1980s I read an article by eminent political scientist John Herz. His concern about our inability to deal with threats to human survival was similar to my own. I sent him something I had written—not really expecting a reply. But I received one. He encouraged me, and we have been in contact ever since. His support was the nourishment the seed needed to sprout, and in time it became this book. Along the way I have received encouragement from others, especially Ervin Laszlo, who also invited me to participate in a meeting of the General Evolution Research Group at the University of Bologna in 1989. Jeffery van Davis, who helped with pre-production planning for a television documentary, *Invisible Walls,* and Kenneth Boulding, who was to appear in it, were also reassuring. (Production funding was unobtainable and filming did not take place.)

Since then I have had helpful discussions with Ralph Cautley, Charles Massion, Jon Moulton, David Ricks, Roger Stephens, Richard Stewart, Donald Swain, and Rudolph Treuman. Henry Felson, Joseph Levee, Scott Mitchell, and David Ricks have kept a supply of pertinent clippings coming my way. Dick Bozian, Henry Felson, Alan Kahn, Daniel Kline, Dan Laycock, Joseph Levee, Peter Lloyd, Frederic Sanborn, and Vern Uchtman have read and commented on parts or most of the manuscript.

David Ricks did some early and final editing. Henry Felson carefully read all the drafts of the chapters as they appeared, providing valuable comments and gleaning out mistakes. And I cannot thank my collaborator, my computer, enough. Without it, I could have never written a book. Besides enabling me to get a grasp on what I was writing, it did the necessary job of storing and recalling information for someone whose memory resembles a sieve more than a file cabinet. It has been a true collaboration of man and machine.

But that still is not enough. My superb editor-agent, Carol Cartaino, with some valuable inputs from Oscar Collier, brought order to an assemblage of ideas, and surgically removed many favorite but too often repeated words and examples. Steven L. Mitchell and Kathy Deyell at Prometheus Books helped bring the book into its final form. And yet one more thing was needed—the help and willingness of my dear wife, Angela, to put up with a state of semi-widowhood as I labored to finish this project.

Index